T0335741

Better Software Project Management: A Primer for Success

Better Software Project Management: A Primer for Success

Marsha D. Lewin
CCP CMC, FIMC

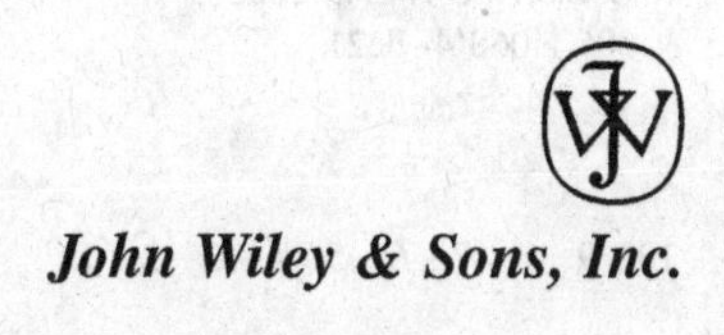

John Wiley & Sons, Inc.

This book is printed on acid-free paper.♾

Published simultaneously in Canada.

Library of Congress Cataloging-in-Publication Data:

Lewin, Marsha D.
Better software project management : a primer for success / Marsha D. Lewin.
p. cm.
ISBN 0-471-39555-2 (cloth : alk. paper)
1. Computer software--Development--Management. I. Title.

QA76.76.D47 L49 2002
005.1′068′4--dc21 2001023797

10 9 8 7 6 5 4 3 2 1

To Forrest, for his love and inspiration

Contents

Preface

Computers may have become commonplace in business, but rare indeed is the successful implementation of computer software. Despite reduction in cost and size of the hardware, software costs continue to run out of control. We may know how to build and install hardware more profitably, but we've not been able to develop software more effectively.

Now that development tools that are easier to use have become available, more people are being called on to develop software with others—but many do not have the management skills for doing so effectively and efficiently. When the fledgling developer is assigned to manage a team, emphasis is typically on getting the software designed and written quickly; he or she is not given adequate time to become the best project manager he or she can become. The initial software development effort provides the baptism by fire that still is the school for software project managers.

That said, after decades of software development projects, there are *some* guidelines that can be imparted—some more quickly than others, of course—to make the process easier, and to increase its chances for success. That's what this book does.

With more than 30 years of software project management experience in private and public sectors, in client/server implementations, as well as in local area networks (LANs) and mainframes, distributed and centralized, I present rules of thumb and techniques for bringing your project in successfully.

WHO THIS BOOK IS FOR

This book addresses specifically the needs of the beginning software project manager, but also well serves:

- Anyone who finds him- or herself responsible for delivering reusable software in his or her company or department.
- Anyone in any department in a business that develops software.
- Anyone interested in increasing the probability of bringing in better software within budget and on time.

This is a book on managing software development—the resources, including people and budget, the quality of the software developed, the cost, and risks. While software development methodologies are referenced, this is not a development method book. You can find many such books referenced in the bibliography at the end of this book. Instead, this book is about managing the folks involved in the development of software, and increasing *your* chances of success. This is a book about the variety of project types, and how *you* can manage them. In particular, this book is geared to managing small- to medium-sized projects, involving the implementation of packaged software (COTS, commercial off-the-shelf) or small development (larger projects typically already have an in-house development methodology and project management approach, with an ISO 2000 requirement).

APPROACHING SOFTWARE PROJECT MANAGEMENT

When I first started in software project management (and for many years thereafter), I found it to be an art, not a science. After all, software development wasn't even regarded as a science then, let alone was the management of the folks who were trying to create it! I am convinced, in retrospect, that the projects I delivered successfully back then were due to my tenacity and to my "people skills." The ability to lead technical people toward a common objective is still an art, although now we have many tools that we can use to represent our planning prettily; but they don't necessarily increase our chances for a successful completion. Nevertheless, I am still convinced that software project management is more art than science, and it's an art that can be learned, should be learned, and it's what this book tries to teach.

I believe there's not just one way to manage a project. A project is, by usual definition, the bringing together of resources over which we often have no direct control, to develop something once. Just as each software project is unique, hence differs from others, we need a set of tools in our managerial toolkit from which we can select according to the needs of a particular project. It's knowing which tool to use—and when—that makes this undertaking an art.

My goal in this book is to give you a large set of tools to put into your own kit. And, hopefully, you'll use them wisely and appropriately as you manage your own projects.

USEFUL FEATURES OF THIS BOOK

The hallmarks of this work are its simplicity and emphasis on teaching how to manage a software project, not just how to wield software tools. While making the process simple, I also impart a conceptual understanding of the principles of good project management, thus avoiding the rote of a cookbook approach. This primer presents more than mere methodology; it addresses the "people" issues that are critical to successful software project management, and identifies ways to reduce risk in *your* project.

Particularly useful features are:

- The first part of the book aims at getting you up to speed quickly, followed by more in-depth background, described in project management terms, after I have established context for you.
- The organizing principle is the need for a software project to satisfy simultaneously the Triple Constraint (schedule, budget, and quality) within the context of risk management (the Quadruple Constraint).
- Subject matter is covered chronologically, from a project's beginning to its end.
- Typical problems are highlighted, to help you sidestep them; potential solutions are presented.
- Dozens of sample forms, formats, and checklists are included to help you visualize what is needed to get started quickly, and to customize for use in managing software projects.
- Different types of software management are addressed, from traditional life cycle to prototyping.
- Different types of software projects and the nuances of managing them successfully are addressed: client/server, LANs, Internet/intranet, mainframe, existing system conversions, packaged software, new and maintenance projects.
- Political issues affecting project success are addressed, as are technical issues.
- Methods are identified for managing your project's risk.

- The role of managing people and their perceptions is addressed.
- Testing and its role in project management are addressed.
- Examples are drawn from diverse projects—contract management, police, airport—to provide variety and breadth of applications (and to prevent boredom!).

HOW THIS BOOK IS ORGANIZED

As I did in my previous book, *The Overnight Consultant* (John Wiley & Sons, Inc., 1995), I have presented a "quick start" in the first part of this book. If you're suddenly thrust into the position of managing others to get software developed, this section will give you what you need right away. Project management in the abstract is far less instructive than using real-life examples.

The next sections give more background in software project management, providing theory behind the quick start, and expand on those issues that make every project unique. Checklists are provided for you to adapt over time and to use in your own projects. Different types of software development methodologies are discussed, as are different types of software projects and the nuances of managing them successfully.

And, as no book is complete without its bibliography, you'll find one here, too, along with a glossary of techno-babble, and an index. We use Microsoft Project, the current industry standard, for explanation and demonstration. In lieu of a list of additional software project management software, I suggest you do a search on the World Wide Web for the latest and greatest products.

Acknowledgments

This book is based upon the approach to project management I learned from my mentor, Milton D. ("Mickey") Rosenau, Jr. Our earlier collaboration produced *Software Project Management: Step by Step* (Van Nostrand Reinhold, 1984, second edition 1988). Mickey's books on successful project management and product development have become bibles in their fields. A complete list (at least as of this writing) of his publications can be found in the bibliography of this book, and I encourage the reader interested in general project management to add those books to his or her library and to visit Mickey's Web site at www.rosenauconsulting.com. My decades of software project management experience are placed within that context.

In the nearly two decades since that collaboration, the software development world has changed greatly: microcomputers dominate the hardware world, and everyone with a copy of Microsoft Access has become a developer. Moreover, everyone with a copy of Microsoft Project thinks he or she can become a project manager by preparing a multicolored Gantt chart! In fact, the proliferation of unsuccessful software projects continues, regardless of the platform, suggesting that more needs to be done to teach good management techniques, rather than instructing in the use of the project management software tools.

I've placed software project management within the context of a methodology with which I've worked for many years—Formula-IT, designed by James E. Kennedy, who specializes in enterprise systems strategy and implementation. This allows me to relate the art of software project management to the science of managing the development in a cogent fashion. I've found the

Formula-IT methodology to be intuitive—easy to learn and quickly grasped by others. It has proven to be scalable to all project sizes, and provides controls that produce desired results according to plan. I encourage you to visit Jim's Web site at www.FormulaIT.com for further information.

I've supplemented Mickey's Triple Constraint framework—budget, schedule, and quality (or performance)—with risk management, which adds a context within which the Triple Constraint must be evaluated. For example, developing software for a brand-new hardware platform using a new release of an operating system is highly risky. One of three things will occur: the operating system and hardware integration will not proceed smoothly, the developed software will not integrate with the operating system, or nothing will work together. Successful strategies for handling such situations will have budget, schedule, and/or quality effects, and you, as the project manager, have to account for those risks within your Triple Constraint framework.

Representing a Quadruple Constraint, a four-dimensional model, in two dimensions is a nontrivial task, so I'll talk about both the Triple and Quadruple Constraints throughout the book, depending upon the best application for the particular situation. You can decide for yourself which construct works best for you.

The question arises: Can you have good software project management without a development methodology? The case can be made that just breaking a job into smaller tasks in a coherent fashion (such as work breakdown structure) is a methodology in itself. In the absence of a methodology, I believe a project can still be well managed. On larger projects, the methodology will bring you closer to the high level of software quality, to which we all aspire.

But this is the place to make acknowledgments, so I shall acknowledge away: to Keith and Jim Kennedy, who have provided not only software development methodology (Formula-IT), but their wisdom and counsel over many years of our collaborative consulting. To Ira Gottfried, who taught me that the best way to manage in certain situations is not to overmanage. To Bill Cordo, who gave me my first software development project, and my baptism by Wall Street's fire. And my special thanks to the hundreds of clients in more than 30 years of consulting who had the confidence in my skills to allow me to manage their software projects and people, and to benefit from the friendships that result from working cheek by jowl (well, e-mail by fax) to achieve the difficult, if not the impossible.

As always, no book can be created without the collaboration of tireless professionals who, even in this age of technology, cannot be replaced. My thanks to my editor at Wiley, Bob Argentieri, who gave me this opportunity to write again on successful software project management, and whose counsel and patience are greatly valued. My thanks also to the Wiley staff, who provided the practical guidance and help that make books out of manuscripts:

to Shannon Egan, my patient managing editor; to Janice Borzendowski, who superbly edited the copy; and to Stacey Rympa, the assistant who followed through the nits and crises. Finally, thank you to my family, to Forrest, my muse; to my children, David, Rozi, Steven, Ron, and Bobbi; to my munchkins, Leah and Sarina; and to my friends, who patiently endured my absences and unavailability as I wrote this.

I hope you gain much from it, that you emerge with a perspective on the task at hand, and a set of tools in your management toolkit that will enable you to keep your projects moving ahead, to a successful outcome. Please visit me at my Web site, www.marshalewin.com, to see what's new, and to ask questions that you may have.

Better Software Project Management: A Primer for Success

1

Getting Moving

As the old joke begins, I have good news and I have bad news; which would you like to hear first? Okay, for you optimists out there, the good news is that because you have done such a good job at writing special database applications for the rest of your contract management department, we've decided to make you the departmental analyst. The bad news is that there will be no increase in pay, no decrease in your concurrent workload, and you will have to deploy a system over the company's wide area network for use by the contracting staff in the remote offices. Oh, and the system has to be ready in six months, for the start of our capital development cycle (with $8 million in contracts to be managed).

Though the type of the system to be implemented may change, this basic scenario is all too familiar in today's environment: someone is made responsible for delivering software developed by others to satisfy the needs of those who may not understand the technical dimension of what they're asking for, nor the business issues that should determine the resources to devote to this effort. If the responsible person is you, you're expected to learn on the job, or through baptism by fire, because, at best, you have had some software development training (well, at least a course in Microsoft Access) but the world of managing software projects is unknown to you.

PROJECT MANAGEMENT QUICK START

What I do in the following pages is present a set of issues to look at, introduce some decisions you will have to make, and, finally, offer some actions you can take to help you successfully bring in your project—ideally, on time, within a realistic budget, and in a fashion that will satisfy your important users. Notice I did not say *all* users—remember, Abe Lincoln warned that you can't satisfy all of the people all of the time, and if he couldn't do it, then surely the rest of us will not be able to either.

For easier exposition, and for clarity, I'll use the example of a contract management system. You can substitute document management system or online purchase requisition system or whatever satisfies your own immediate needs. One word of caution here: I am not talking about a multiyear development effort with dozens of people. For that, you'll need more than this quick start section: you'll probably need a supply of Tylenol, and to read this entire book. Remember, this part of the book is intended to get you going quickly, and to increase your chances of success *now,* when you need it most. The remainder of the book explores in greater depth the basic theory you'll need to master if you decide to stay in software project management, and expands with examples, checklists, and forms. Of course, your successful first endeavor will probably determine whether project management is a proper use of your energies—and if you have the talent for managing software projects.

Sinking or Swimming

So, what's a project manager to do to get started? First, don't panic. I had a friend who was diagnosed with a dreadful disease, yet when we spoke about it she was in fine fettle. She was handling her disease as she would any other project management task, by taking charge. She was:

- Learning about the project.
- Identifying any constraints.
- Determining what the alternatives were.
- Setting the goals to be achieved.
- Prioritizing the goals.

Her basic goal was to get better, understandably, but she had to decide among many treatments, and that process was confusing. Once she weighed her alternatives and established the primary goal, she then scheduled her treatments and executed her plan. Budget was not the issue, although for some in her situation cost might impact the treatment goal, specifically by limiting the treatment options.

Which brings me to the point of how this analogy relates to *your* project management background. Software projects rarely have a single user, a single way to achieve the goal, a fixed amount of money to achieve the goal, and a set time frame. If you're in that situation, then you're among the few fortunate ones indeed. Usually, a project has a set of choices that have to be made, and, like unwinding a terribly snarled piece of cord, everything is intertwined, so that making a choice in one area has consequences in others.

That choice dilemma is the background for the *Triple Constraint.* Mickey Rosenau taught me years ago that a project has three dimensions that must be satisfied, and they are not the same for every project.[1] These dimensions are the limits, or constraints, that define your project and help determine how you're going to execute your project.[2] These are:

- Quality of the software, of what you're implementing (performance)
- Cost to implement (budget)
- Length of time to implement (schedule)

Figure 1-1 shows where different project types would fall along the constraint axes.

If your project involved, say, a space launch, then clearly performance would be the most important. If you had to shoot when Mars was in its proper, but briefly accessible, orbit, then schedule would also be important. However, if the software wasn't ready in time, you wouldn't go ahead and launch anyway, and hope for the best, because, while budget was less important than performance and schedule, it still would be important—and quite possibly not replaceable.

[1]Cf *Bibliography,* especially Mickey Rosenau's *Successful Project Management.*

[2]According to the *Project Management Institute's A Guide to the Project Management Body of Knowledge* (PM BOK Guide), 2000 edition, Newtown Square, PA, Project Management Institute, p. 43, a constraint is a factor that limits your options, such as contract amount, time, and the specifics of what you are developing.

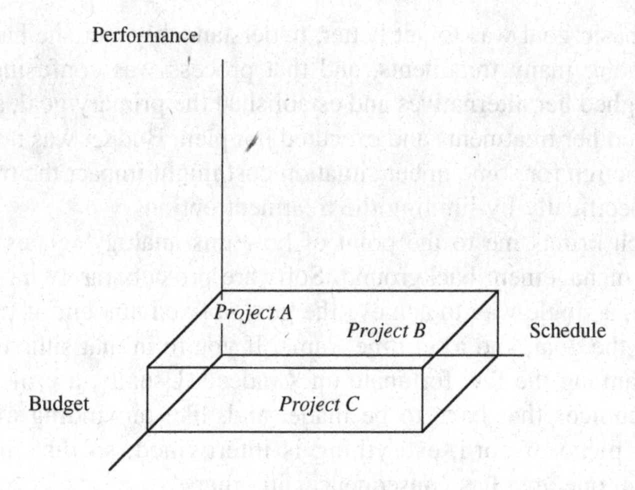

Figure 1-1 The Triple Constraint and the placement of projects within.

On the other hand, if you are developing software for tracking delegate activities and needs at the Democratic National Convention, your major focus in the Triple Constraint would be schedule, because the software will probably not be current enough in four years' time for the next convention!

In many businesses, budget is the primary constraint. This is especially true in certain public sector projects, where only a certain amount of funding is available. You must go back to boards of supervisors or commissioners for additional funding, which never goes well. When budget is the most important of the constraints, you'll need to accommodate the constraint by changing the performance, for example, by producing less than originally planned with the project, or, in rare cases, by using ancillary resources on a limited basis, and taking longer to deliver the software. (This is generally not such a good choice, as taking longer generally costs more.)

The point here is that if you are responsible for managing a software project, you need to know what's important to the successful outcome of this project, and how the constraints rank against one another.

To accommodate doing business in this new millennium, I've added a fourth constraint to this situation, risk: How much risk can you tolerate within the project? For example, recent software projects focused on Y2K issues.[3] The element of risk posed by older software, which

[3]Year 2000, when older software might have failed because it couldn't accommodate the date change to the new millennium.

would otherwise have had no constraints, made these projects so important that they were budgeted and scheduled for completion before the millennium bells chimed. Budget, on the other hand, was not a primary concern. An estimated $320 billion were spent to reduce the risk of companies and governments being unable to operate on January 1, 2000.[4]

Thus, because addressing risk is so important in my view, when I talk about planning, it is in the context of the *Quadruple Constraint.* However, when I address monitoring, I will continue to use the Triple Constraint, because risk issues are manifested in the budget, schedule, and performance. Both constraints are addressed in greater detail in Chapter 2, but they come into play with your contract management project.

What Are *You* Trying to Do?

It always improves your chances of success if you understand what you're supposed to be doing. Now that you're called a software project manager, what does that mean for you? And while I'm at it, let me raise here some easy-to-grasp but necessary concepts that should help you to put your arms around the project and to increase your chances of a successful outcome (which is what this is all about, right?).

We begin with the question, what is software project management? Your task, should you decide to take it on, is to manage a one-of-a-kind effort in the field of software development or enhancement. Projects are, by definition, unique undertakings, with definable starts and measurable ends. They differ from ongoing functional tasks, say as an accounts payable clerk's repetitive cyclical processing of checks due vendors, or human resources' hiring of personnel. In these two examples, while the payee will change, or the position being filled will vary, the basic function does not change.

Software project management is uniquely different from other management tasks in that you don't see the final result until much later, and you use tools that might not have been previously tried. This latter fact is due to a constantly evolving technology, which both gives us new ways of doing things and new ways of tripping ourselves up. Additionally, you are dealing with users who, often, have never dealt with software implementation before, which also means you are

[4]See Dewayne Lehman, "Senate: Y2K Fixes Worth the Billions Spent," *Computerworld* online, March 1, 2000, at www.computerworld.com/cwi/story/0,1199,NAV47_ST041669,00.html

dealing with their perceptions of what you're bringing to them, since they cannot know beforehand its impact. Unlike hardware, which is more concrete—there is something to touch and feel—it is typically harder to test "soft" specifications of processes that users will have to incorporate.

As unique undertakings with nonrepetitive goals, software management projects present some unusual situations not found in other types of endeavors. Here are brief introductions to a few of the issues you need to address (you'll find more complete descriptions in Chapter 2).

Methodology and Management

There are many existing approaches to the actual development, or building, of software. These are called *methodologies*. This is *not* a book on methodology, which is the process of building software; this is a book on the management of the development process, which I refer to as *software project management*. Common methodologies include life cycle, or *waterfall*, models, which are the traditional way of delivering software. An example of this model can be found in Figure 1-2. Individual steps occur sequentially, from feasibility assessment of the project, to final installation and acceptance of the software.

As microcomputers became commonplace, users wanted more involvement with the development process earlier on, raising the demand for prototyping, or evolutionary delivery. This allowed users to see and use pieces of the software prior to complete delivery, so that they could offer constructive feedback, thus making it possible to incorporate changes cost-effectively into the software. An analogy in the construction industry is design-build, where a design is left open for completion after construction starts, instead of waiting for the detail of the entire design to be completed before the first shovel is put into the ground.[5]

The management of the software development project then ties into various milestones and monitoring events within the methodology. However, if your company had a methodology, you'd probably be given a manual with the description of how to use the method, and the

[5]There are many other types of development methodologies. Felix Redmill's *Software Projects* (John Wiley & Sons, Inc., 1997) is a good source for a more complete description of the distinctions between software development methodologies, software engineering, and software project management.

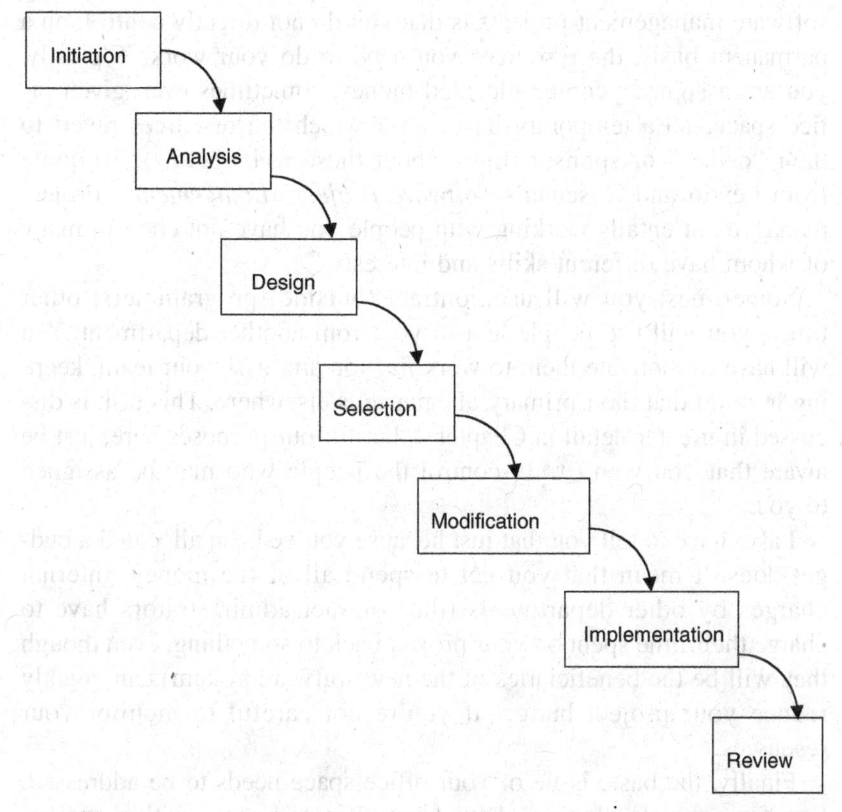

Figure 1-2 *Software development methodology: Waterfall life cycle model.*

automated tools that go with it. You may well need to learn how to manage rather than wield a tool, but if so, you probably wouldn't be reading this section of this book right now.

The point is that the names and sequences of the activities are determined by your methodology, and it is the management of that methodology that we're addressing here. So, for those of you who have neither method nor background, let's get on with it. Suffice it to say that you'll decide which method of development to use as part of your planning process.

Identifying Your Resources

I'm sorry to be the one to have to tell you this, but you won't be in control of the resources you need. But a unique characteristic of software management projects is that you do not directly control, on a permanent basis, the resources you need to do your work. Typically, you are assigned people, allocated money, sometimes even given office space, on a temporary basis, after which the resources revert to their "owner" or sponsor (more about these folks below). To quote from Lewin and Rosenau's *Software Project Management,* "Project management entails working with people you have not chosen, many of whom have different skills and interests."[6]

Sometimes, you will use contract (outside) programmers; other times, you will use people lent to you from another department. You will have to motivate them to work *for* you and *with* your team, keeping in mind that their primary allegiance is elsewhere. This task is discussed in greater detail in Chapter 4, but for our purposes here, just be aware that you won't truly control the people who may be assigned to you.

I also have to tell you that just because you've been allocated a budget doesn't mean that you get to spend all of the money. Internal charges by other departments (the contract administrators have to charge their time spent on your project back to something, even though they will be the beneficiaries of the new software system) can quickly reduce your project budget if you're not careful to monitor your resources.

Finally, the basic issue of your office space needs to be addressed. Say, for example, that you have been allocated space within another department, or you are piggybacking on other projects. What happens if they are compelled to relocate? You may well find yourself and your staff spending time packing, unpacking, and squatting at others' desks during critical project times—and without the extra time to do so, thereby reducing your available resources further.

WHAT'S A PROJECT MANAGER TO DO?

Now that I've painted this unpleasant scenario, I want to quickly reduce the anxiety of having to develop software to satisfy a group of users, and using resources you don't control. You are, simply put, going to *manage.*

[6]Lewin and Rosenau, *Software Project Management: Step by Step,* 2nd edition (Marsha D. Lewin Associates, Inc., Los Angeles, CA, 1988), p. 7.

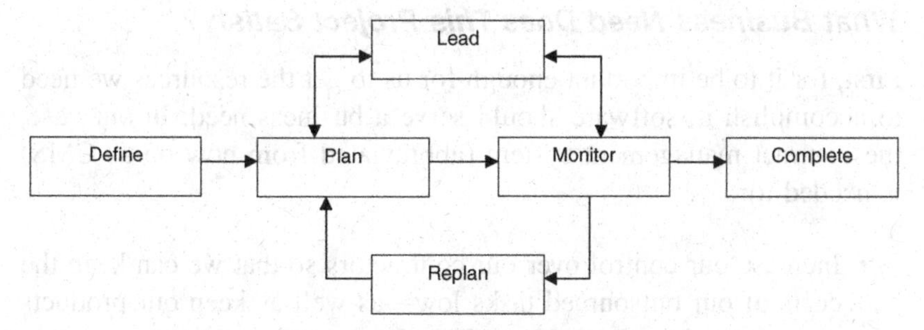

Figure 1-3 *The project management process.*

Project management, whether of software, software and hardware, or my friend's health, is a simple, five-step process, steps that generally overlap:[7]

1. Define.
2. Plan.
3. Lead.
4. Monitor.
5. Complete.

Figure 1-3 shows the steps in this process.

Because this is a quick start section, I'll leave the details of each of these five steps until later in the book. In this part, I'll give you just enough information to get you through the first project, which in this example is the contract management project.

Define

What are the goals of this project? One of the major problems in getting a project going is that often the goals are not the same for everyone. As soon as you have more than one user, you have multiple goals—which isn't bad as long as they're not at opposite ends of the spectrum, which they often are. Answers to the following questions will help you decide what to do when you're in the proverbial swamp: to play with the alligators or to drain the swamp!

[7]Lewin and Rosenau, *Software Project Management: Step by Step*, 2nd edition, (Marsha D. Lewin Associates, Inc., Los Angeles, CA, 1988), pp. 7–8.

What Business Need Does This Project Satisfy?

First, for it to be important enough for us to get the resources we need to accomplish it, software should serve a business need. In our case, the contract management system (abbreviated from now on as CMS) is needed to:

- Increase our control over our contractors so that we can keep the costs of our outsourced tasks low—as well as keep our products competitively priced. Note: The company intends to outsource more of its processes, and will have more contracts to manage.
- Reduce the overpayments being made by our company on its prior contracts.
- Allow us to compare among contractors to determine which keep closest to budget so we know which to hire.

Once we have a well-established business need, then our project will be seen as important enough to get people to spend time to help define and test the system, and to accept change.

What Other Needs Does This Project Satisfy?

In addition to business needs, we may have to meet various other specific needs:

- The department head wants uniform processing of all contracts so that the staff is interchangeable among contract administrators.
- Uniform processing reduces training costs, and executive reporting becomes easier when all contracts are being managed using the same software. Subsequently, they can be rolled up more easily to satisfy "quick questions from on-high."
- The accounting department wants more timely information. Currently, the contractors are getting paid too slowly after they submit their invoices, and the company is having trouble finding contractors willing to do business with it.
- A new accounting system is going in, and the contractor payment requests must dovetail with the new system.
- The Information Technology (IT) department wants Windows software to run on its Novell wide area network so that it can reduce its support costs.

And so on and so forth. Note that I have not included an individual employee's desire to learn, for example, SQL Server, as a requirement. It might be nice to have, but it definitely is not a requirement.

How Do You Rank the Needs?

If we have *measurable* needs, so much the better. How much money do we lose by inefficiently processing contractors? How many contracts do we have in force? Since, as just noted, the company is going to be outsourcing more in the future, how many contracts does it anticipate having? How many contract administrators does it now have, and how many contracts can each administrator manage? How many contract forms does the company have? How will this project be paid for? Who/which department(s) will pay for this project? I think you get the picture.

The point here is to quantify as many of the requirements as we can. This will separate the wish list from the basic needs. When we get into the planning stage, we will have some firmer numbers to work with that will help us rank our goals.

Plan

It really is much easier to plan where we're going when we know where we're going. The question then becomes, do we know where we are going? What are we planning? This is where the Quadruple Constraint comes in. We need a framework, a way to lay the foundation, then build around it, to give order to what we want to do.

Schedule Constraint

I always start with the schedule constraint, but others may choose to start with cost, performance, or risk. In our case, we need to know when the new accounting systems are coming in or when the company plans to start increasing its outsourcing. That information obviously will limit some of the options we might plan for. Let's say we have one year before the number of contracts is going to double. That means we have a year to acquire or develop, test, implement, and fully train our staff before the onslaught of new work. Table 1-1 shows the schedule issues associated with our CMS project's goals.

Next we analyze the goals against cost; this is where the quantification helps. For example, it turns out that the company overran

TABLE 1-1 Schedule Issues of Project Goals

Project Goal	Schedule Issues
Increase control over contractors.	Reach agreement on critical control factors.
Reduce overpayments.	Identify payment process and decide how to reduce steps in process.
Compare contractors.	Start gathering comparative data immediately to identify trends; apply results.
Reduce costs of managing contracts.	Implement earlier to save more money.
Process contracts uniformly.	Meet legal department requirement that contract forms be reduced within six months.
Reduce training costs.	Plan outsourcing of personnel for one year from now.
Facilitate executive reporting.	Reap political benefit of faster implementation.
Supply more timely information to accounting department.	Produce more accurate reports that reflect activities within monitored time frames.
Pay contractors more promptly.	Recognize that projects will start sooner and potentially end more quickly.
Implement new accounting system.	Prepare test plans for pay requests in new system.
Run software on Novell network.	Obtain software certification for Novell, and test on network early enough to identify any configuration changes needed.

$100,000 last year on the 250 contracts—totaling $4 million—that it managed. That's an average of 2.5 percent overrun, which is anticipated to be reduced by the new system. If we double the number of contracts, as planned, for a total of 500 contracts, without a new system, we have the potential for an overrun of $200,000. This can, in short order, reduce the company's viability. (Brings to mind the old joke: If we can't make a profit on one item, we'll make up for it in volume!)

Now look at Tables 1-2, 1-3, and 1-4. They add to the schedule issue of our project's goals the issues of cost, quality, and risk, respectively. Note that not all factors have issues associated with them. Those that do may have issues associated with the particular project's content, while others are related to the Quadruple Constraint.

If we stand to lose $200,000 without a system, we should be able to spend that $200,000 on the new system, and be in a more controlled situation. So now we see a budget figure emerging. For now, we won't add in the costs of personnel savings or facilities, to keep this simple. Now that we've come up with a broad-brush schedule of a year, and a potential budget of $200,000, let's look at another constraint dimension: performance.

TABLE 1-2 Cost Issues of Project Goals

Project Goal	Schedule Issues	Cost Issues
Increase control over contractors.	Reach agreement on critical control factors.	Write new procedures, as needed, into existing contracts; consult legal counsel for this.
Reduce overpayments.	Identify payment process and decide how to reduce steps in process.	Identify labor costs to handle new processes versus old methods.
Compare contractors.	Start gathering comparative data immediately to identify trends; apply results.	Determine whether data can be gathered readily, or budget will have to include manual research of old files.
Reduce costs of managing contracts.	Implement earlier to save more money.	Determine which costs to include in these figures.
Process contracts uniformly.	Meet legal department requirement that contract forms be reduced within six months.	Acquire additional consulting assistance as needed to determine legality of agreements and contract forms.
Reduce training costs.	Plan outsourcing of personnel for one.year from now.	Develop new training materials.
Facilitate executive reporting.	Reap political benefit of faster implementation.	Take into account that drill-down methods of preparing executive reports may require more expensive database software.
Supply more timely information to accounting department.	Produce more accurate reports that reflect activities within monitored time frames.	Hire additional clerical help as required to verify accuracy of reports, and for possible development of interim automated verification tool.
Pay contractors more promptly.	Recognize that projects will start sooner and potentially end more quickly.	Gain executive consensus for cash flow implications to company.
Implement new accounting system.	Prepare test plans for pay requests in new system.	Note that costs of this project may reflect development of additional interfaces (to existing and new systems).
Run software on Novell network.	Obtain software certification for Novell, and test on network early enough to identify any configuration changes needed.	Budget for additional testing.

TABLE 1-3 Quality Issues of Project Goals

Project Goal	Schedule Issues	Cost Issues	Quality Issues
Increase control over contractors.	Reach agreement on critical control factors.	Write new procedures, as needed, into existing contracts; consult legal counsel.	Allow for long verification period to assure correctness.
Reduce overpayments.	Identify payment process and decide how to reduce steps in process.	Identify labor costs to handle new processes versus old methods.	Establish criteria early, especially for partial payments and inquiries.
Compare contractors.	Start gathering comparative data immediately to identify trends; apply results.	Determine whether data can be gathered readily, or budget will have to include manual research of old files.	Ensure that criteria are objective; record corporate experience.
Reduce costs of managing contracts.	Implement earlier to save more money.	Determine which costs to include in these figures.	Define activities that comprise contract management, and determine which are to be maintained in new system.
Process contracts uniformly.	Meet legal department requirements that contract forms be reduced within six months.	Acquire additional consulting assistance as needed to determine legality of agreements and contract forms.	If unusual project-specific terms require too many different contracts, perhaps agree to uniform clauses instead.

Reduce training costs.	Plan outsourcing of personnel for one year from now.	Develop new training materials.	Set up focus groups to identify best training topics and methods.
Facilitate executive reporting.	Recap political benefit of faster implementation.	Take into account that drill-down methods of preparing executive reports may require more expensive database software.	To ensure uniformity of information across divisions, hire personnel or implement systems as required.
Supply more timely information to accounting department.	Produce more accurate reports that reflect activities within monitored time frames.	Hire additional clerical help as required to verify accuracy of reports, and for possible development of interim automated verification tool.	If information audit before release deters promptness, allow for additional testing and possible policy decision.
Pay contractors more promptly.	Recognize that projects will start sooner and potentially end more quickly.	Gain executive consensus for cash flow implications to company.	Uncover errors in evaluation of work quality before making payment.
Implement new accounting system.	Prepare test plans for pay requests in new system.	Note that costs of this project may reflect development of additional interfaces to existing and new systems.	If all features cannot be implemented completely, implement in phases.
Run software on Novell network.	Obtain software certification for Novell, and test on network early enough to identify any configuration changes needed.	Budget for additional testing.	Verify that related software now performs without problems and security levels are correct.

TABLE 1-4 Risk Issues of Project Goals

Project Goal	Schedule Issues	Cost Issues	Quality Issues	Risk Issues
Increase control over contractors.	Reach agreement on critical control factors.	Write new procedures, as needed, into existing contracts; consult legal counsel.	Allow for long verification period to assure correctness.	Recognize that some contractors will not do business with us.
Reduce overpayments.	Identify payment process and decide how to reduce steps in process.	Identify labor costs to handle new processes versus old methods.	Establish criteria early, especially for partial payments and inquiries.	Check that new processes don't introduce errors.
Compare contractors.	Start gathering comparative data immediately to identify trends and apply results.	Determine whether data can be gathered readily, or budget will have to include manual research of old files.	Ensure that criteria are objective; record corporate experience.	Be aware that subjective assessments may distort reality, and older data may be unusable.
Reduce costs of managing contracts.	Implement earlier to save more money.	Determine which costs to include in these figures.	Define activities that comprise contract management, and determine which are to be maintained in new system.	Acknowledge that new procedures may be hard to learn, and errors will increase in the short term.
Process contracts uniformly.	Meet legal department requirement that contract forms be reduced within six months.	Acquire additional consulting assistance as needed to determine legality of agreements and contract forms.	If unusual project-specific terms require too many different contracts, perhaps agree to uniform clauses instead.	Accept that perceived rigidity of terms of contract may deter some contractors from doing business with us.
Reduce training costs.	Plan outsourcing of personnel for one year from now.	Develop new training materials.	Set up focus groups to identify best training topics and methods.	Note that when training is inadequate, more errors may be introduced.

Facilitate executive reporting.	Reap political benefit of faster implementation.	Take into account that drill-down methods of preparing executive reports may require more expensive database software.	To ensure uniformity of information across divisions, hire personnel or implement systems as required.	Safeguard against varying quality of underlying information, which will result in wrong decisions.
Supply more timely information to accounting department.	Produce more accurate reports that reflect activities within monitored time frames.	Hire additional clerical help to verify accuracy of reports, and for possible development of interim automated verification tool.	If information audit before release deters promptness, allow for additional testing and possible policy decision.	Beware that faster information may not be accurate, causing errors to increase.
Pay contractors more promptly.	Recognize that projects will start sooner and potentially end more quickly.	Gain executive concensus for cash flow implications to company.	Uncover errors in evaluation of work quality before making payment.	Note that vendors and others may apply pressure to implement this component earlier, increasing project risk.
Implement new accounting system.	Prepare test plans for pay requests in new system.	Note that costs of this project may reflect development of additional interfaces to existing and new systems.	If all features cannot be implemented completely, implement in phases.	Guard against possible loss of control and issuance of payments in error.
Run software on Novell network.	Obtain software certification for Novell, and test on network early enough to identify any configuration changes needed.	Budget for additional testing.	Verify that related software now performs without problems and that security levels are correct.	Maintain software to run properly, and preclude network shut downs.

Performance Constraint

Every project has a performance goal, even if it's not specified. Typically, when processing financial information, we aim for 100 percent accuracy on the amount specified in the reports. The columnar representation will vary according to whether it's an internal report (where we don't print in colors) or a report going to clients and customers (in which case we'll want balanced columns, attractive type fonts, maybe even logos). And on invoices, if we produce a tear-off stub to return with payment, it helps to have all the identifying information located on the stub so our accounts receivable clerks can properly process the payment!

Clearly, the performance constraint has many dimensions. How quickly can the program prepare the information? For example, in the good old days (that is, pre-PC), it took hours for the stock brokerage industry to process its daily trades. This created situations where today's trades might not be ready for the next day's business. As computers got faster, this problem went away (fortunately for today's Internet-based home traders). But this business sector would not have prospered if speed could not have been accommodated. For obvious reasons, in the financial industry, accuracy produced after the fact would not have been good enough in our software development.

The plan lays out what we're going to accomplish, how long it will take to complete it, and how much we anticipate it will cost. Because writing things out helps not only to communicate to others, but to consolidate our own thoughts coherently, this plan should be in writing (or e-writing, using contemporary technology).

In our CMS, for example, the plan might look like the one shown in Appendix I. It:

- Describes the goals of the project.
- Defines the performance the CMS will deliver, in more detailed terms.
- Indicates how long it will take.
- Estimates what it will cost, including dollars, reflecting the mix of human and physical resources needed.

Additionally, because graphics help to illustrate the information contained in lists of numbers and words, the plan should be accompanied by a schedule chart showing how long the project will take, and, perhaps, a management presentation summarizing the critical points we want the reader to grasp. These are shown in Figure 1-4 and Text Box 1-1, respectively. We'll discuss how to accomplish these later, but for the moment realize that, today, we have available many computerized tools to deliver these materials in professional-looking formats. But first come the analysis and planning; these tools will later reflect the good management process underneath.

Management Constraints

The Quadruple Constraint affects out planning process—how we'll slice and dice the project to achieve success on the four dimensions that will define our ultimate success. If no risk is the major goal and we can find outside vendors with successful records of implementation of the CMS for clients with our size and technology environment, then we're in luck. We'll simply outsource the project, and our role will be to manage the external, rather than an internal, staff, for development.

That said, keep in mind that even with external contractors, we'll still need to work with our internal users and management in the steps that follow.

Lead

As soon as we begin to talk about leading, we enter the human realm of a project: leadership of the individuals, sometimes those within our own company, other times those working for outside contractors, to achieve the project goals. One of the most difficult management tasks is to take the assembled ragtag group with diverse interests and get them to work toward a common goal for a limited period, after which they'll return to their own departments or to another project. Knowing the project is temporary, some will always just sit by and wait for the project to end, offering minimal commitment. As a consultant, I face this consistently. The good news is that most people *want* to do a good job. Often, however, they don't understand what they're expected to deliver, or even what the definition of "good" is for this project. And, in software projects especially, the intangible nature of much of the development process exacerbates misunderstandings and ambiguities, which confuse expectations.

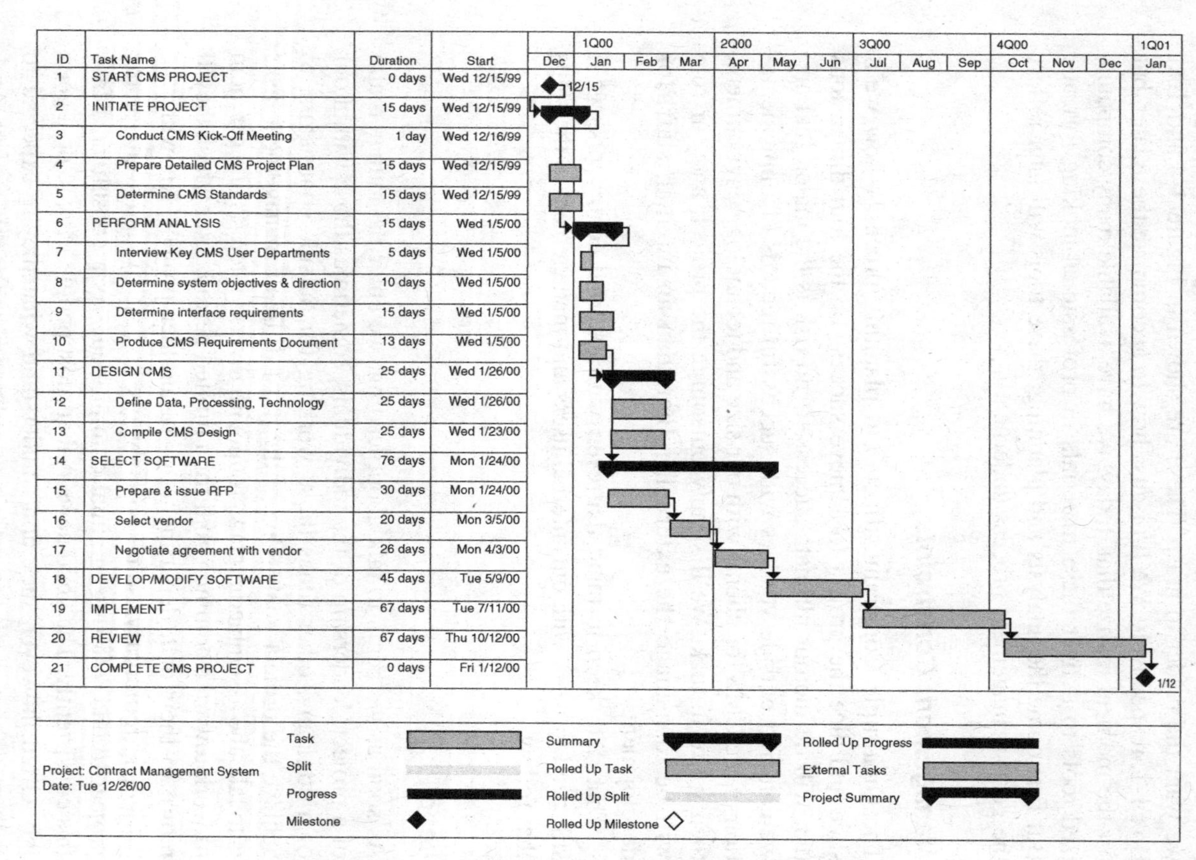

Figure 1-4 *Contract Management System Time Line.*

TEXT BOX 1-1 CONTRACT MANAGEMENT SYSTEM MANAGEMENT PRESENTATION

What Is It?

- Common method of managing contract documents and costs companywide

Advantages

- Can easily roll up costs across all departments and divisions.
- Can easily compare projects.
- Common training reduces costs.

Project Goals

- Control outsourced tasks to keep costs low.
- Reduce overpayments to contractors.
- Identify best-performing contractors and repeat-hire those.
- Enable executive roll up of all project information without manual keying.

Acquisition Strategy

- Buy rather than develop.
- Evaluate proven products available off the shelf.
- Develop internal training capability.

Anticipated Return on Investment (ROI)

- Current costs of staff to handle inquiries on late payments
- Costs to retrain staff due to high turnover
- Eighty percent reduction in above costs anticipated.

Conclusion

Though leadership philosophies and styles vary, the common thread running through well-led projects is that the individuals involved become a team, committed to a positive outcome, one on which each member feels accountable for performance. The project manager only guides the team to achieve the outcome; he or she does *not* do all the work for the team.

In our CMS, who comprises the team? Identifying the constituents is often the hardest part of the project. Wherever people are involved, a project becomes highly vulnerable. Great strides have been made in the technological aspect of project management, but the weakest link remains human interactions.

Gather the Team

In our CMS, the obvious team members come from the user and the technical communities—the contract administrators and the programmers. Another member is the management sponsor, who can help move things along faster. The sponsor is typically the executive, or committee of executives, charged with ensuring the project's success through adequate funding and resource allocation. The sponsor approves the project definition. (See Appendixes I and II for further detail.) Perhaps a representative from the finance or accounting department would also be of help, especially if the planned-for system will feed into such systems.

This brings up two points regarding the teams we form:

- The team should comprise as wide a group as possible.
- The team composition will change over time as the project develops.

When my team works with our clients, we usually introduce the team at the start of the project to those who have a stake in the outcome. Those include downstream users of the system, as well as the day-to-day users, who will benefit from the efficiencies such systems will confer. (After all, why bring in a new, *less* efficient, system?) When people who may be affected by the new system aren't included, we run the risk of not finding out how other systems will be affected. What good is a system in one part of a company that requires another system in the same company to become more inefficient in order to accommodate it?

At project launch, your team will typically be composed of many users. During the course of the project, some will drop out, because they leave the company or because their priorities prevent them from attending meetings or being heavily involved. It is essential that the management sponsor attend the first meeting, to kick things off—because he or she can give your project the formal company blessing that will make things easier as you proceed. After that, if the sponsor cannot attend, it is imperative to keep him or her informed of the team's progress.

Though the team will decide what needs to be done, they will need guidance from, for example, technical people as to the technological consequences of the alternatives being suggested and discussed. Let's assume, for instance, that someone knows of a CMS that runs on a UNIX system, but that your company is an NT shop. That raises the question: Who will provide the assistance to integrate with your existing systems? Not only can that cost more money than you have budgeted, but you may not be able to acquire the people to complete the task on schedule. Also, you might end up having to program interfaces between your system and the NT-based financial system.

When you start to program, you'll probably need more technical people at meetings; conversely, at this stage, the users will not be needed so often. Once the system is ready for testing, however, the users will again be required to see how well the system satisfies the goals originally set. A nice complete circle, if you will. A sample project team composition is shown in Figure 1-5; the changing nature of

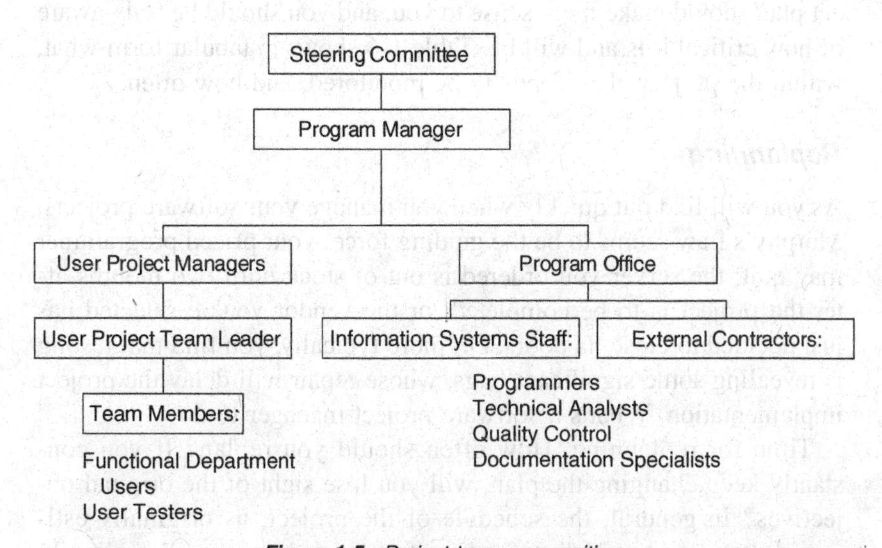

Figure 1-5 *Project team composition.*

the members is shown in Table 1-5. Chapter 4 has more on leadership issues.

Monitor

The most important part of project management is to make sure that everything is proceeding on time and within budget, to achieve the project goals. Monitoring may, by some, be misconstrued as nagging, but it's a necessary process for determining where both people and events are in the plan.

To monitor properly, you need to plan properly: specifically, that means completely and unambiguously. Otherwise, how can you expect to accurately measure where you are? I emphasize this point, because so often projects fall behind because the monitoring isn't done properly—every week the number of lines of code developed are dutifully counted, and the team attends scheduled meetings, but the functionality isn't checked out for the seemingly simple reason that no one included the testing step in the project plan. The team will be in for an unpleasant surprise if they test everything at once and find out it doesn't work!

Chapter 5 includes a more comprehensive checklist regarding what you need to monitor, but, for our purposes here, I identify a number of items that your plan should specify to monitor. At this stage, the project plan should make more sense to you, and you should be fully aware of how critical it is and will be. Table 1-6 shows in tabular form what, within the project plan, needs to be monitored, and how often.

Replanning

As you will find out quickly when you manage your software projects, Murphy's Law seems to be the guiding force: your prized programmer may quit; the server you ordered is out of stock until two months after the project is to be completed; or the vendor you've selected has just decided to close its doors. Or, more typically, you find that testing is revealing some significant bugs, whose repair will delay the project implementation. What's a software project manager to do?

Time for replanning. How often should you replan? If you constantly keep changing the plan, will you lose sight of the original objectives? In general, the schedule of the project, as originally estimated, is a good guideline for how frequently you should replan. In

TABLE 1-5 Project Team Involvement by Life Cycle Stage

Team Members	Initiate	Analyze	Design	Select	Modify	Implement	Review
Senior Management	x			x			x
Functional Dept. Representatives	x	x	x	x	x	x	x
Main Users	x	x	x	x	x	x	x
Technical Analysts		x		x		x	x
Programmers					x	x	x
Quality Control Analysts		x			x	x	x
User Testers						x	x
Documentation Specialists					x	x	x
Project Manager	x	x	x	x	x	x	x

TABLE 1-6 Monitoring the Project Plan

Task	Frequency	Reason
Check on availability of resources.	At start of each project phase	Key personnel may be involved in continuing work, or have left company; cash flow may be less than anticipated, reducing funding.
Check on technology infrastructure (network, hardware, and systems software).	Throughout project	Unforeseen infrastructure changes may require changes in your project.
Verify user availability.	Especially during vendor selection, training, and testing	Management often is unaware of the time commitment necessary at these points.
Get authorization to proceed.	During and at end of contract/agreement negotiations	Legal resources often are not immediately available, and can hold up project start if agreement is not signed and approved.
Acquire hardware.	As needed	Hardware procurement policies often introduce delay.
Establish test and acceptance criteria.	Prior to implementation	Without adequate criteria definition, implementation may be inadequate and not detected until too late.

your project plan, you will have built in milestones, or checkpoints, when you will decide whether changes in schedule, budget, or performance are required. As a guideline, I suggest the following, but be aware that in your environment, you may find other frequencies that work better.

> *Note:*
> Another method of replanning is to do it when you reach the end of each phase in a project. This is detailed in Chapter 5.

As a rule of thumb, I try to keep replanning to a minimum; and if the project is under 90 days in duration, you probably will not require any replanning at all. As the length of the project increases, the number of checkpoints increases, such that you might want to replan every six months, or annually on a multiyear project.

Our CMS is only a six-month effort, so I would suggest monitoring biweekly (every other week). I would monitor costs as frequently as invoices are paid by the company, or when labor hours are charged. For example, if labor hours are charged to the project weekly, I recommend checking weekly. That will tell you if people are spending more time or less on a given effort, and you'll have enough time to go back to them and have them make up any deficits—or to identify why more hours than planned are being expended. If, on the other hand, invoices are being paid semimonthly, a weekly review of the cost reports won't be of great help, since there'll be no change every other week.

If the system is supposed to go into full production ("go live") within the month, and users are still finding bugs in programs, that's a good time to replan the implementation. Clearly, though, there is no single rule for how often to check; you'll need to develop a "feel" for frequency after you've managed a few projects, and after you've dealt with different types of people and situations. If you are going to err on one side of caution, I suggest erring on the side of monitoring too frequently.

Complete

All project managers hope their projects will be completed and approved by the stakeholders on time, so that they can move on to the next project. But how do we define completion? It's over when it's over? Not exactly. From the point of view of the people who have to maintain the software, they are just beginning when you think you're done. So we're back to the project definition. There, you should have defined not only the goals, but the scope of the project: Does it encompass the entire life of the software? Or, when the design and initial implementation are completed, does it move over to the information systems (IS) department, where changes to the software are managed as a separate project? And in your project plan, did you define what would constitute *acceptance* of the software? What good is running a system for two weeks without errors if the software is designed to work with month-end closings, and the two-week period covered is in the middle of the month?

These questions bring us back to the project plan and statement of work. The project plan represents the formal, approved document used to guide the project execution and project control—indicating how it will be done, what resources are needed to do it, and how it will be accomplished. The *statement of work* (SOW), on the other hand, describes the work to be supplied to the project to achieve its objectives.

The plan should have spelled out the acceptance criteria—acceptance by all the stakeholders in your project—at the outset. But, since your knowledge will be greater at the end of the project, the detail of the plan need not be provided at the beginning. But one of the project's milestone events that takes place before the programming even starts, while the users are deciding what they want, is to establish the test and acceptance plan.

The project plan will then identify that acceptance comprises the successful completion of that plan, to remove the ambiguity caused by not clearly specifying the criteria. At the end of the project, participants are generally in a hurry to finish up and start a new project, or perhaps go on that overdue vacation; external contractors want to get paid, and be free to work with other clients. By not defining, or too loosely defining acceptance (if you've left it until the end), the result may be software that is unsatisfactory or difficult to maintain, satisfying no one.

Acknowledging Completion

I strongly recommend that you get acknowledgment of your project's completion. This can be handled at a final team meeting, with consensual sign-off by the members. Or you can have each member fill out an assessment report. You may also want to couple the sign-off with a "lessons learned" meeting, a review of the entire project, wherein you identify, and document, things that you might do better next time, or future enhancements to the system. The folks responsible for maintaining the software may be able to implement suggestions from this process.

For example, let's say the CMS doesn't roll up estimated costs into the companywide budget system. The lessons learned meeting reveals that a planned release of the vendor's software would allow for export of those numbers from foreign systems (the budget system) and comparison against the CMS figures. Discrepancies are reported—in other words, a debugging tool will be provided. This will reduce maintenance time, as well as testing time, when implementing enhancements. That will be valuable for planning subsequent maintenance projects.

More commonly, with software projects, mandatory requirements often reflect the way people perform their work under the existing systems. The new system typically introduces more efficient ways of performing tasks. A lessons learned meeting might identify features originally requested that weren't really necessary, or procedures that can now be streamlined. In the old system for example, requests for payments due under the contract may have had to be reviewed individually by a supervisor prior to being approved, then forwarded to the finance department for payment. Under the new system, the supervisor can query all contracts and determine which require approval—all at one time. Or, instead of having a written transmittal to the finance department, the approved requests for payment can be forwarded to the finance department electronically with an accompanying e-mail form.

Go Nitpicking

In 35 years of managing software projects, I've never been involved with a project that didn't have a few little "nits" that still needed attending to upon completion. In construction, they call the list of things to be done at completion a "punch list." Throughout the project, as you monitor, it's a good idea to compile your own nit list, to make sure actions don't fall through the cracks. At the end, no doubt you'll still have a few "close-out" items, such as final documentation, transmitting electronic files to the records management department, and preparing a list of contacts for the maintenance staff.

Minor items on the nit list should not prevent you from completing the project—just make sure you don't include milestones (such as successful acceptance test!) on your list. Remember, the nit list should contain *only* minor items that someone else can see to and that do not affect the completion status of your project.

MATERIALS REVIEW

I covered a lot, quickly, in this chapter, so let me summarize the material as a brief review. Your task is to manage the development of software. To do that requires taking five interdependent managerial steps, often simultaneously, though they are sequential in nature:

1. Define.
2. Plan.
3. Lead.
4. Monitor.
5. Complete.

The concept of the Quadruple Constraint serves as the framework to help you plan your project, which constitutes: specification of performance, schedule, budget, and risk. The statement of work in Appendix II defines the project's milestones. How you will accomplish the project is defined in the project plan, as shown in Appendix I, while the methodology defines the technical steps required to accomplish the software development. Your project management will monitor how well you're proceeding using the methodology to complete the project.

You will be faced with many challenges as you work to complete your project on time, within the budget, and to meet promised functionality. You may have to alter your original plans to complete the project, but the changes will still have to satisfy the Quadruple Constraint in order to have the project deemed successful. You should document any changes to plans, and obtain any approvals necessary to ensure you really have completed the project.

2

Mastering the Process

As promised in Chapter 1, the following chapters of the book go into greater detail on topics covered quickly there. Presumably now you've got a frame of reference both from the mistakes you made and successes you achieved during your initial project management foray; and hopefully, ideas presented earlier are starting to make more sense. But you also have more questions now. This part of the book will answer them.

Here, the five interdependent activities of a project are further broken down and described more fully. Other issues of particular importance in software projects are highlighted as well. Examples of relevant forms and documents are provided for you to build on. Starter checklists are included for you to add to (remember, a checklist is an individual tool—you have to adapt it to your particular situation).

Onward!

THEORY OF CONSTRAINTS: THE DIMENSIONS

All types of projects have constraints, or are limited by many things—personnel and financial resources, quality, and time, among others. But I maintain that on software development projects, the risk constraint is of great importance because software, by its very ethereal nature is riskier business. I identify risk as a separate constraint because projects will not be launched if their risk is determined to exceed an acceptable level. Therefore, identifying that level, and building in the safeguards on the other dimensions of the Triple Constraint, is a critical part of the planning process.

Monitoring risk is typically done along the three dimensions of quality, budget, and schedule. There are formal monitoring documents (test results, cost reports, and project schedule) to reflect the progress, and risk reports refer back to the three report types. So, for purposes of planning, I refer to the Quadruple Constraint when I'm addressing planning issues, and to the Triple Constraint when discussing monitoring of software projects. The reason for doing so is because I want

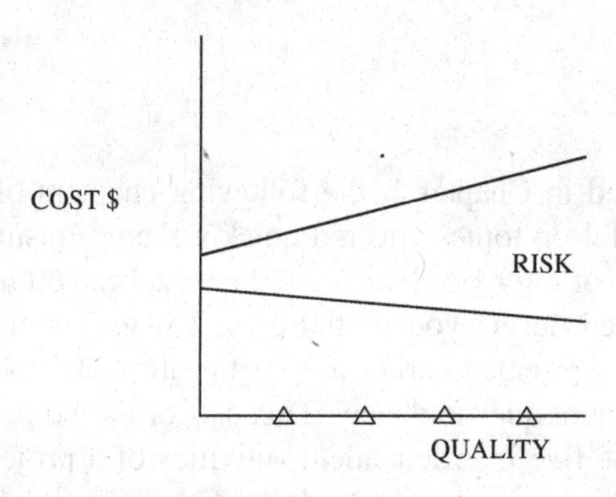

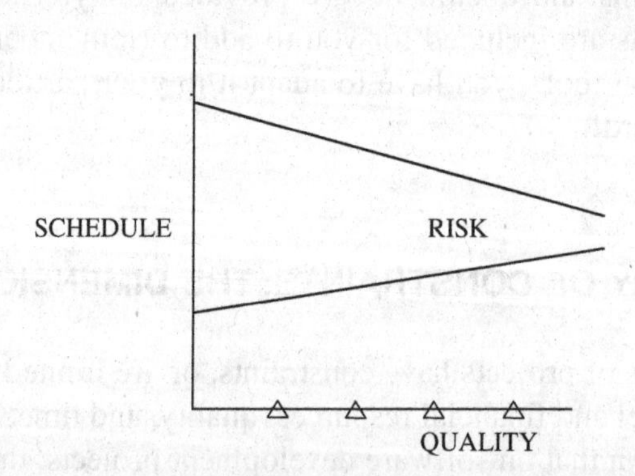

Figure 2-1 *The Quadruple Constraint. Quality is measured in milestone achievement (indicated by △) on the X-axis. These examples assume that the particular software project has increasing risk as cost increases, while risk decreases as schedule lengthens. However, a particular project may have exactly the opposite situation.*

to convey to you, the reader, the nature of defining and planning projects along important dimensions.

Software projects can be defined, uniquely, on the three dimensions of cost, schedule, and performance, as shown earlier in Figure 1-1. When adding the fourth dimension of risk, graphical representation becomes more difficult. I attempt to do so in Figure 2-1, by breaking risk into two three-dimensional representations. The point here is that you will alter your risk, sometimes adversely, when making cost and schedule decisions.

By determining what you'll deliver, and for how much, and when it will be ready, you establish a baseline against which your subsequent project management decisions can be measured. To repeat, successfully managing a project means delivering what you promised within the budget and schedule to which you've agreed, without introducing unacceptable risk to the project.

Now, that sounds fairly concise and easy to understand. Unfortunately, satisfying a project's Quadruple Constraint is very difficult, because the events that naturally occur during the lifetime of a project conspire to lower performance below the specification, and to drag the project behind schedule, which usually makes it exceed the budget and raise the risk of failure.

SATISFYING THE QUADRUPLE CONSTRAINT

You, as the project manager, must stay alert to the problems, and strive constantly to satisfy the Quadruple Constraint, if you are to successfully manage your project. And it is this book's purpose to make you a better manager, so let's discuss some of the obstacles you may encounter and the problems you may face.

I organize them according to the dimension they impact most, although some affect all dimensions (these are, needless to say, the worst kind). For example, if you spend more time testing because the code was not developed properly, you'll not only impact your schedule, but you'll probably overrun your budget due to the increased cost of hiring additional personnel to conduct more testing and do the reprogramming.

Addressing Performance Problems

Probably the most difficult dimension to achieve is the performance specification. Software professionals have spent decades trying to perfect this, without success. Indeed, forests have been felled to make paper on which to print the millions of specifications that have been written but that have missed the mark.

Cause: Poor Communication

Poor, imprecise communication is one cause of this repeated failure. Conveying a concept in words is, by nature, imprecise: the words and figures we choose to explain to another what we mean may instead cause confusion because each of us has a different perspective; hence, our words and illustrations mean one thing to us, and only later do we discover they meant something entirely different to the user. For example, let's assume a user wants a simple data entry screen, to allow for maximum speed in the inputting process; the supervisor, on the other hand, wants field-level validation, to ensure that nothing erroneous enters the system. The specification developed reads: "easy to use," an inherently ambiguous phrase that will result in disappointment for one or the other party—a simple, rapid entry screen will necessarily become more complicated, and take longer to use, if the supervisor's needs are also designed into the system.

Cause: Too Much Communication

Too much communication is another problem, causing much to get lost in the entropic process. Software specifications in any form are merely an *attempt* to define as unambiguously as possible an implementation that will be written by someone else, usually for use by yet another party or parties. It is inherently difficult to define as-yet unseen, unused software to satisfy current and future user requirements. Ambiguity creeps in whenever anyone else, for example a programmer, implements a "best guess" of what was meant in the specification.

Cause: Technological Changes

A third problem arises because of technological changes—and their implications. For example, in this era of frequent releases of new versions of software, a project charged with upgrading an accounting package may be fraught with unforeseen peril: hardware limitations may prevent the new version from performing quickly enough; the software may cause the network to lock up; and so on. It is rare indeed these days that a workstation does not connect to a network—local area or the Internet—and thus is affected by factors not apparent to the user.

Cause: Poor Programming

A fourth problem is caused by poor programming. Software development tools have certainly improved over the years, but the abilities of programmers have not necessarily kept pace. And even the best programmers can, and do, make mistakes. The resultant software may have performance deficiencies that prevent achievement of the Quadruple Constraint.

Cause: Variable Situations

Finally, it is essential to point out that performance is multidimensional. Your software must conform to company standards and interface requirements, both of which impose constraints on your software's performance. Satisfying everyone is not always—if ever—possible. Therefore, compromises need to be worked out, so that the performance delivered is acceptable to all project constituents. For example, in using the company-mandated database management system or servers, a one-second response time may not be achievable for all types of activities: you may be able to validate within that time, but a query on an entire database will require 30 seconds. The point is, the specification must clarify the response time achievable in different situations.

Identifying Time Problems

Scheduling problems arise for several reasons. The most common one I've discovered is being pressured to meet a delivery date imposed by external sources, usually management, without, from the outset, having adequate resources to meet that date.

Cause: Overemphasis on Performance

The most insidious cause of scheduling problems is an overemphasis on the performance dimension, at the expense of a balanced view of the Quadruple Constraint. For instance, computer scientists (who are frequently appointed as project managers) tend to concentrate on the technology, and to strive for technological innovations or breakthroughs. Many are the computer programmers who have spent time unnecessarily working out a clever algorithm, or wielding a new programming language, rather than simply patching an existing program.

This preference for so-called elegant solutions, rather than for practical implementation, is met at the expense of the schedule, and frequently is accompanied by unfavorable cost repercussions and increased risk. As is often said, "better" becomes the enemy of "good enough."

Technically trained people often treat the performance specification as sacrosanct, while believing the schedule and cost dimensions can be altered. They'd rather their work be judged as late or expensive, than as inadequate, quality-wise. They also feel that accomplishing the impossible, regardless of risk, is notable.

Cause: Resource Unavailability

A second reason for scheduling difficulties is that necessary resources are unavailable when required. Unavailable resources may be equipment or personnel. For example, in a recent implementation, needed workstations and servers for deployment were three months late, because the vendor did not have the inventory to meet a promised delivery date. We had no choice but to slip the schedule because other alternatives, such as using temporary platforms, would have resulted in double conversion work and higher labor costs.

The point is that the unavailability of planned resources forces the project manager to accept substitute solutions, which may require outsourcing the programming. Or, worse, it may mean using marginally qualified people, who will take longer to do the work and make more mistakes than the well-qualified technician initially counted on to be available. *Technical people are not interchangeable,* not even in the same labor category—we've all known analysts who could do the work of three other analysts combined. Being very specific as to the resource's qualifications—or name—will increase the probability of getting that resource assigned to your project.

Cause: Disinterest

Third, sometimes a project encounters scheduling difficulty because those assigned to it are not interested in their tasks. In some cases, they may choose to work on other tasks, or to work half-heartedly on your project. This problem is particularly acute in projects that already have extended well past their original schedule, which usually forces personnel to work on crash schedules for months at a time.

There is a limit to the adrenaline flow that enables programmers to work around the clock in the hopes of finally completing the project. A manager must stay alert to signs of burn-out in team members, especially those who have been working on the problem for too long. Generally, their productivity lags, and their errors increase in number and severity.

Cause: Personnel Changes

Fourth, scheduling problems often occur because of personnel changes over the life of the project. Any project over six months in duration can count on personnel changes—at the very least, in the user community, if not among your team and/or management. This introduces two risks; the aforementioned resource unavailability, and the task of bringing onboard and up to speed a new technician or manager who must assimilate the terminology, task, and personalities involved in the project.

Cause: Scope Creep

Fifth, you'll notice schedule slippage when the performance specification is raised—often referred to as "creeping scope." This is common when features over and above the original specification are added to the original project—often in answer to the users' delight at what they see. Infectious enthusiasm often results in well-meant efforts to add a few "trivial" bells and whistles. If you give in to this, keep in mind this means you're agreeing to do additional work, without altering schedule or cost. Certainly, no single item will put the project way behind schedule, but it doesn't take many such changes to produce a one-day schedule slippage, a one-week slippage, and so on until the project is seriously behind schedule.

Facing Cost Problems

Cost problems arise for many reasons, often related to performance and schedule problems. When a project is in trouble on its time dimension, it often reflects in the cost as well, because resources are not being used as efficiently as planned. If software is not performing according to specification, further unforeseen costs for additional hardware, such as another server, and software, such as an operating system or database, may be required to fix the problem.

Cause: Negotiating Games

A second cause of cost problems is the "liars' contest" that often occurs during contract negotiation, if the project is being done for an outside organization. When the organization is mandated to accept the lowest bid, this is particularly prevalent. Let's say you bid $500,000 to fully implement a maintenance management system for a client. During the negotiations, you are told to "sharpen your pencils," by at least 10 percent, or they will award the work to another vendor. In your desire to win the project, you and your managers agree to minor wording changes, indicating a modest reduction in the scope of work, but with a substantial cost reduction. When you reduce price without fundamental work reductions, you have built in a cost overrun at the very outset of your project. An experienced project manager will never agree to this kind of negotiation unless he or she knows the money will be restored in later contract changes or additions to the scope. But for the new project manager, this is a potential cost problem to be aware of.

The liars' contest also occurs internally. When a project must be sold to upper management or to other divisions, you are competing with other managers for authorization of a project. In fact, in some companies and public agencies, you'll find that management requires internal departments to compete with external resources on software development projects. The assignment is awarded to the best qualified, and is outsourced or insourced appropriately. You can figure pretty quickly that if an IS department loses too often to outside sources, it soon will be only a maintenance shop—and even that can be outsourced!

Whether you're involved in the liars' contest internally or externally, I recommend that you pretend you're dealing with an external customer in all cases, and go through the rigor of defining and planning. Actually, considering all the internal politics in many companies, a strong argument can be made that formality is even more important on internal assignments, where you'll probably have ongoing contact with your "customer" and others who lost the project to you!

Cause: Unrealistic and Inestimable Cost Estimates

A third source of cost difficulty is due to estimating initial costs too optimistically. These estimates simply don't reflect general practice: inefficiencies that occur when scheduling resources to perform the work; substitution of equipment at the projected price, due to unavailability, with more expensive replacements; and labor rate increases.

In certain situations cost problems arise because the work is so new that the project manager—you!—cannot be sure what will be required. You may know which hardware components and systems software and software tools to assemble, but are unsure what you'll need when you actually implement. I was involved in such a case recently. We were switching from an SNA architecture-based service to a TCP/IP service. The best-laid plans didn't prevent us from needing both types of connections until the service completed its conversion—much later than scheduled. Our budget obviously reflected the double installation and monthly maintenance costs, not to mention hardware expenses.

Reduction of risk is often reflected in cost—for additional layers of testing, for hiring outside services. The secret is to budget for acceptable risk, rather than eliminate all chances of risk. That is, however, impossible in some cases, such as the recent Y2K experience proved. In general, you can automate 80 percent of a desired project scope more effectively than the entire scope. Sometimes the last 20 percent costs 80 percent of the entire project, and the complicated nature of the resultant software increases the risk to unacceptable levels.

Cause: Human Error

Not even the best project manager is perfect, so sometimes costs estimates are wrong simply due to mistakes. Even in this era of automated spreadsheet calculations, operator error creeps in—vital columns are dropped, or the wrong data is used. For example, in justifying the cost of a new records system, and trying to show cost savings over current methods, charging the wrong costs for labor will show an overrun very quickly.

Cause: Poor Management

Another reason for cost problems that is inexcusable, but that nevertheless occurs, especially in smaller projects that are internal to a department, is poor management: the project manager is not cost-conscious, or, worse, does not have an adequate cost management system. With so many tools now available—even a simple automated spreadsheet can suffice—this should never happen.

Cause: Funding Failure

Sixth, anticipated funding may not be forthcoming according to plan. This is never a good thing, for a well-planned project has a cash flow against which the project has been structured. If funds are not available at anticipated points in the project—say, when hardware is to be acquired, or additional labor costs are anticipated for testing and training—rescheduling of the project will be required. And that, as already discussed, can cause further cost and schedule delays. For a project to stay within its budget, not only is funding needed, but it is needed as scheduled.

Cause: Phased Implementation

When a project is to be implemented in phases, larger total project budgets are generally needed. Because resources are being allocated phase by phase, without knowing whether people and facilities will be needed for the total length of time, negotiation for them must reoccur at each phase. The resources may cost more in the new phase; additional personnel may have to be trained, to replace those who have left the project; and certain software features may be necessary for a particular phase but not for permanent use.

Addressing Risk Problems

Logically, risk increases substantially as problems occur on one or more of the other dimensions. And because software development is an inherently risky business to start with, you may find that the project is no longer feasible because your sponsor does not want to subscribe to the increased risk.

Cause: Too Many People

A common problem is using too many people on a given task, resulting in a greater chance of miscommunications as software modules are integrated. And those with less development experience may produce code that is not up to par. This can be avoided by using fewer, but more experienced and highly qualified, developers.

Cause: Poor Integration

A second cause of risk problems is inadequate correlation of requirements and testing, to ensure that what has been asked for has, in fact, been delivered. This can be relieved by using a software development methodology that incorporates multilevel testing back to the original requirements.

Cause: Schedule Slips

Risk increases when a project schedule slips. And once a schedule slips, you run the risk that the quality of software developed will be lower, due to inadequate testing caused by time pressures. Also, the original users who assisted in defining the system may have left the company by the time you deliver, and the new users may have new definitions and may try to alter the deliverable or, even worse, refuse to accept it at all.

TAKING CORRECTIVE STEPS

With the Quadruple Constraint as your planning guide, you can eliminate many potential problems, and mitigate the effect of others. When you develop software you follow principles that govern the design (for example, speed of data entry is more important than retrieval speed). Likewise, in planning your project, you will have greater success when you follow planning, monitoring, leading, and completion principles. That's what the Quadruple Constraint helps you to do.

Step: Solicit User Support

Users are critical to your project's success, so soliciting their support in helping you meet the Quadruple Constraint is an important step. In fact, it's best to regard this step as mandatory. Systems cannot be forced on others; systems forced upon others are ultimately failures. To paraphrase an old adage, you can lead a user to the system, but you can't make the user enter data properly, or think like the system. It's far easier to make the system reflect the way the user can think!

Users can help you satisfy the Quadruple Constraint in a number of ways. Here are some that have been successful on my projects:

- Participate in project teams throughout the life of the project.
- Appoint a user to head the team.
- Commit to the project's success.
- Participate fully in defining requirements.
- Budget adequately, both in dollars and hours.
- Provide adequate clerical and user resources.
- Be receptive to change.

- Know what exists, and clear up any ambiguities.
- Know what is wanted, and clear up any ambiguities.
- Turn around documents quickly, with feedback where needed.
- Establish measurable acceptance criteria.
- Deliver test data, and evaluate test results promptly.
- Adhere to contract.
- Control change requests.
- Clearly set priorities.
- Agree to principles that will guide the system development.
- Agree to principles that will guide the project management.
- Be advocates of the new system, selling it to the rest of the user community.
- Be patient and tolerant.
- Above all, make sure you plan and deliver what the sponsor and users expect.

Step: Produce a Thorough Performance Specification

You can also increase your chances of satisfying the Quadruple Constraint by making sure that the performance specification is well-written and complete. The performance specification will generally be composed of, and relate to, many different documents, as explained in greater detail in Chapter 3. Wherever the information resides, however, and whatever you call the document(s), you must be sure that the following criteria are met when you specify performance:

- Complete, as to all functions to be performed.
- Complete, as to all documentation to be furnished.
- Complete, as to legal requirements.
- Unambiguous, as to meaning and requirements.
- Consistent with data inputs and outputs.
- Feasible and measurable.
- Comprehensible, by users, testers, and trainers, as well as by developers and managers.
- Comprehensible, by individuals who will ultimately maintain the system.

On some projects, it will be impossible to define all the specifications completely. For example, recently, when my team was implementing an enterprisewide geographical information system, we knew that, ultimately, we wanted departments throughout the organization to be able to view common base information, and to develop maps using their own department-specific information. However, we could not specify up front which department would implement its own system, and what departmental information it would need and provide. So we broke the project into phases, each one with a complete performance specification appropriate to that phase. The first phase allowed the departments to envision what they needed after seeing the developed baseline information. The final phase specification was then able to provide measurable detail that could be properly tested. The lesson here is, when you are unable to set all project specifications up front, as happens with longer or more exploratory projects, use phased implementations.

ADJUSTING TO PROJECT OUTCOMES

It hardly need be said that, rarely, will everything go exactly as planned. Similarly, not every project will satisfy precisely the Quadruple Constraint. Generally, there is deviation within the ranges, as illustrated earlier in Figure 2-1.

As always, the devil is in the detail; just how much deviation is acceptable depends upon your specific project's ranking of the four objectives, or constraints, which is why each project is unique. You can, for example, tolerate schedule slippage when accuracy and/or risk cannot be compromised—you'd wait before installing a bug-ridden police dispatch system because lives depend upon it. You can tolerate cost overruns when accuracy cannot be compromised or when a schedule absolutely must be met. In the former, you'd work as long as it took to make sure a new banking system accurately posted customer transactions; in the latter, you'd do whatever necessary to ensure the sensing software for your space launch was ready for liftoff.

Just remember, it's the relative importance of each constraint dimension, unique for each project, that will determine whether your project succeeds or fails.

Essential Points

At the start of any project, failure to clarify the natural ambiguity will lead to failure. Early and ongoing communication between the project manager and the project constituencies will help to eliminate ambiguities and to create a common vision. It will also better enable you to rank the constraints.

The project manager must, at all times, serve as a balance between the technical personnel's desire to perfect beyond schedule and cost constraints, and management's requirements to conserve costs, control risk, and produce something rapidly. Writing clear performance specifications is the only way to ascertain that you've done what you set out to do. You can't measure software against vaporware.

DEFINING PROJECT TYPES

Software projects can be distinguished by *size,* by *intended use,* and/or *complexity*. The type will often affect relative importance along the Quadruple Constraint. For example, a software project might be considered small if it is being done by a single person. On the other hand, I know of a two-year project being done by a single consultant, for a multimillion dollar division with 70 employees. Though I don't consider that small, it's being treated as such by the sponsor. (I might add that it has not been successful, either.)

My guidelines for defining a project type are:

- If the software is to be sold to others, it's a big project.
- If the software is to be used by others across divisions or large departments in an enterprise, it's a big project.
- If the software is to be used internal to a department, it's a medium-sized project—unless there are multiple groups, in which case it's a big project.
- If the software is to be used only by a group, and will be completed within six months with one to two people working on it, it's small; or, if more than two people are working on it, it's medium-size.
- If the software is to be used by a single person, will be completed within a month, and is to be developed by a single person, it's not even accurate to call it a project; obviously, it's small.

- If the software has a high level of risk, it is considered a step larger regardless of the other risk-sizing factors.
- If the software gives competitive advantage, it's larger by one level than otherwise.

Small projects generally require less management because there are fewer resources to coordinate. Medium projects lend themselves to what is covered in this book. Large projects tend to have additional unique characteristics, not only because technology becomes perishable the longer it takes to develop the software, but also because resource availability becomes riskier. Many large projects without a methodology can be helped by the lessons put forward in this book, as resource planning is better than none at all!

Perils of Small Projects

The problems unique to small projects include: tight schedules, tight budgets, small teams, and often, low-priority status. Typically, the costs of administration and project management will consume a bigger proportion of the budget, as the small project needs to be monitored even more carefully.

Also, getting up to speed on any project takes time; this factor is not directly proportional to the size of the project. For example, if a program is to be implemented in eight weeks, and the programmer delays two weeks, the project will quickly be 25 percent late, and possibly overrun the budget by a more significant amount than if the project were 20 weeks in duration. Moreover, because small projects typically have less funding, they also have less contingency from which overruns can be accommodated.

Another hindrance to teams working on small projects is that they generally have access to content experts only on a part-time basis. As anyone who has waited for the Oracle guru to consult for two hours knows, this is often an exercise in frustration. Small projects are often too low on the technology totem pole to get the attention they need. The project manager, therefore, must learn to effectively lobby for such resources. Often, specifying full days, rather than hours, will expedite the availability, but at greater cost for the project.

Studies have shown that high-priority projects are more likely than low-priority projects to be completed successfully, because they will normally win any competition for key resources.[1] You can obviate the problem by keeping the decision makers aware of your project and its importance—or by avoiding managing small projects in the first place. If a project is important enough to earn early commitment, you should strive to obtain your resources before sponsor enthusiasm wanes.

On one project I recall, the project manager failed to request training resources until he was absolutely sure the programming would be completed. The project was delayed each time while we waited in the trainers' "queue." Had the project manager prereserved a date, the resource would have been set aside, and our training could have occurred earlier, while other engagements were reshuffled.

Managing software projects of any size comprises a number of tasks that must be performed: definition, planning, monitoring, leading, and concluding. These tasks will generally take a larger proportion of the total time on a small project. Therefore, you, the project manager, should streamline reporting techniques (for example, instead of a weekly progress meeting, you may want to use e-mail because the team is small, and progress can be reported briefly, in a paragraph). You may issue the project schedule only once, rather than at the end of each phase in a larger project.

The point is, you shouldn't spend more time managing the project than developing the software itself. On the other hand, you must still be firm about documenting and testing, because no project, no matter how small, is immune from bugs and problems. And you must still train, and interface with other system efforts in an organization. I am reminded of a client who had many small departmental software projects underway. The resulting systems generally were not well documented, and many died when the current user left because there was no cross-training. Those systems that did survive eventually fell into disuse because the interfaces among them proved too cumbersome. The company had wasted its investment in the software.

In short, small projects can backfire if the risk is not managed. When this happens, it can serve to decrease user confidence in future

[1]Lewin and Rosenau, *Software Project Management: Step by Step,* 2nd edition, (Marsha D. Lewin Associates, Inc., Los Angeles, CA, 1988), p. 289.

efforts. Finally, small projects can grow into large projects, in which case more formal methods should be used.

Platform- and Architecture-Specific Projects

Platform (hardware and operating system) and architecture (distributed, centralized, client/server) decisions are often treated as part of the design; conversely, projects are often categorized by either or both of these design attributes. How many times did you hear Y2K projects classified as COBOL projects, which indicates the language of the application, but gives no hint as to the application? Likewise, describing a project as a client/server project indicates the complexity of the architecture (multiple applications on the individual workstations with programs and data on the servers), but it does not functionally describe the project.

The architecture and platform describe what you are trying to implement, and can be misleading or inadequate in conveying the business nature of your project, so it's best to use descriptive project names that are meaningful to both management and end users, not just to technical people; for example, an "online purchase order system," or a "document management system."

Once you've selected a design architecture, there are important implications for project management, especially in the testing and integration stages of the project. You will have to allocate more time for testing systems that have multiple instances (distributed, client/server), since failure can happen in any link within the system. Integrating multiple architectures within the same system raises the risk level because you increase at the same time the number of connectivities (each of which requires testing). Networked systems introduce their own complexity, in that the protocols among servers and firewalls must be carefully tested along with the hardware and software.

APPLYING METHODOLOGIES TO THE MADNESS

This is a book on managing software projects; its lessons can be applied to all projects, regardless of how the software will be developed (the method). A development methodology comprises a model of how the development will proceed, and includes design, testing, and

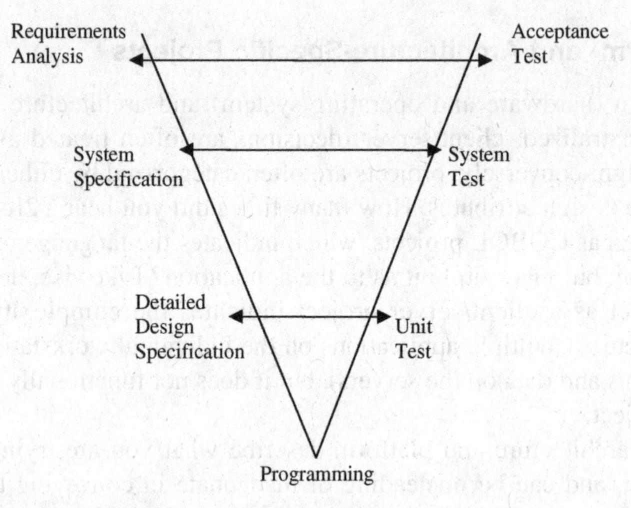

Figure 2-2 *"V" software development methodology.*

completion, as well as a manner, or method, of verifying that the process is proceeding as it should. The most common models are:[2]

- The Waterfall model, shown earlier in Figure 1-2. The steps in the design process are viewed as overlapping and iterative, much like the steps in our project management.
- The "V" model, shown in Figure 2-2. Here, the Waterfall model steps are rearranged, with the conceptual (design) parts on the left, matching up to the testing and validation components on the right.
- The Spiral model, shown in Figure 2-3, is Boehm's risk-based model. It allows for review at each step of the project to determine the best method at the end of each step, or quadrant, to minimize risk. Each cycle goes through each of the four quadrants noted.

The Life Cycle Management methodology is used to manage all of these project models. What they have in common is that they feature

[2]See Chapter 2 of Felix Redmill's book *Software Projects* (John Wiley & Sons, Inc., 1997) for an in-depth presentation of this topic.

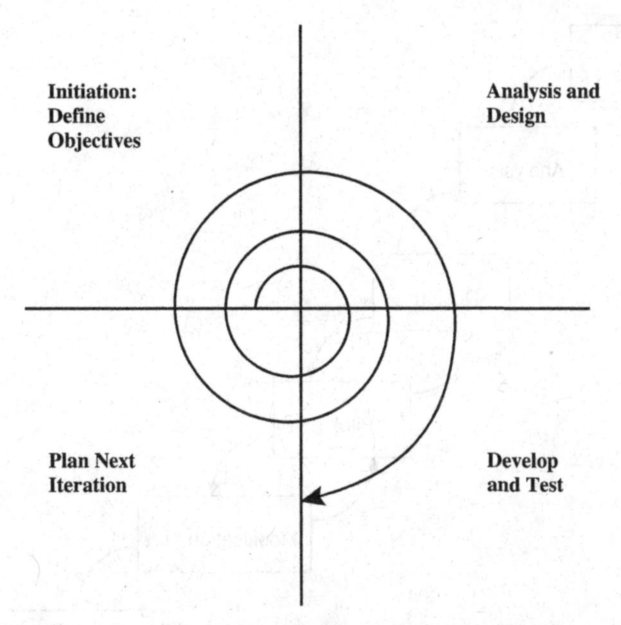

Figure 2-3 *Spiral model software development methodology.*

stages of the software building process; and the methodology incorporates review points at the various stages, to ensure that the project is proceeding properly. This is appropriately called *staged delivery*.

Another methodology, Rapid Application Development (RAD), is popular in the delivery of user-based systems. Design continues throughout the project. RAD is very useful when there is ambiguity in the initial requirements, or when the requirements keep changing. But in order to monitor how well you're proceeding in a RAD method of development, you will have to test differently.[3]

Building a prototype in any of these models is particularly helpful when there is to be a heavy user-program interface, because users approach problems differently from technicians. For example, in a recent implementation, the users preferred keying more information into a single screen rather than invoking multiple screens. This knowledge saved much programming that had been planned.

In conclusion, you want to select a measurable model that best suits your particular goals. I still use a staged life cycle model, with user

[3]See William E. Perry's *Effective Methods for Software Testing,* 2nd edition (John Wiley & Sons, Inc., 2000) for a complete treatment of software testing.

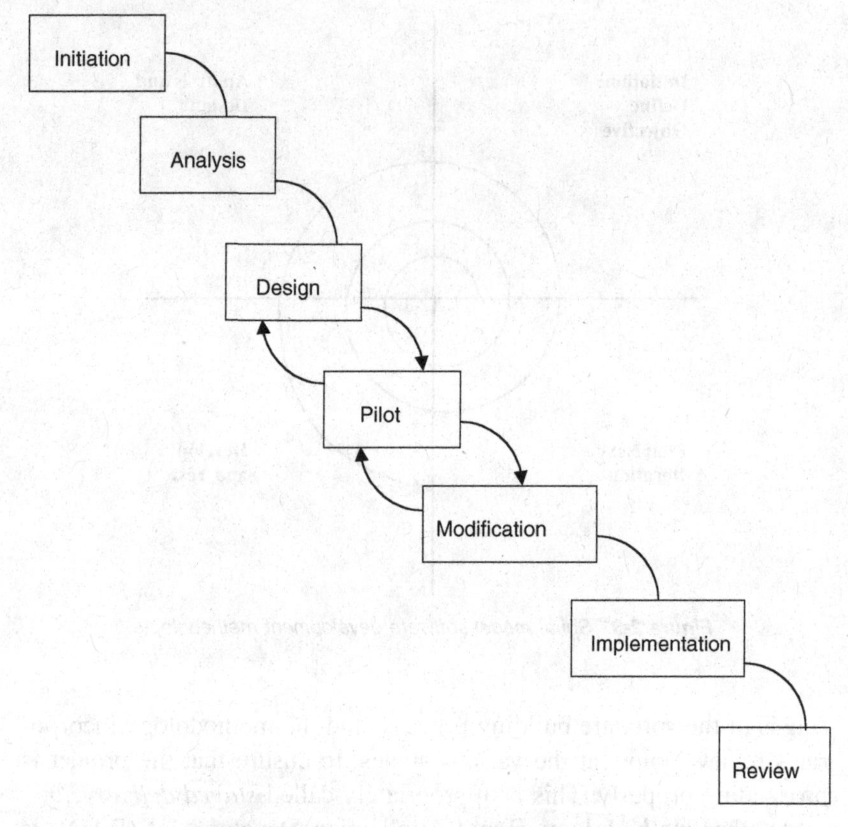

Figure 2-4 *Formula-IT development methodology.*

involvement accounted for in a prototype or pilot phase, as shown in Figure 2-4, Formula-IT.

GETTING THE LAY OF THE LAND: KICK-OFF

Before jumping into specifying what the software will do—which is a natural tendency if you've entered project management from the computer technical ranks—ponder for a moment the *business purpose* your software is intended to serve. Doing so will enable you to determine where your sponsorship should come from, how much support you'll need for this endeavor, and where on the Quadruple Constraint your project will fall. Often, you'll learn that things are not as they seem.

For example, I worked on a software modification project whose goal was to upgrade a human resources department's software, to include an online benefits registration capability. While middle management supported this change, and the employees had been lobbying for such a feature for years, the human resources director was not in favor of it. Through a series of discussions with the director prior to the definition of the project, my team was able to not only reassure him that existing data would remain secure, but that he would benefit from the availability of such information and help him to reduce his backlog of paper forms. In planning the project, we added a pilot program, and solicited his participation.

On another project, my team was charged with automating the processes of a police force. This was to include cellular entry of reports from patrol cars. Upon study, we found that the officers wanted to select their own equipment, and that the mechanics wanted to be part of the team because they had to install the equipment. We had anticipated the officer involvement, but neglected to include the mechanics at first pass. Subsequently, we incorporated them on the project team, resulting in greater efficiencies in the installation and maintenance of the system.

The lessons here should be obvious: look both strategically and tactically for people and situations that can contribute to your project's success; incorporate those people into your teams, and accommodate the situations in your planning.

Playing Politics

Fortunately, most projects I've worked on have been successes. On those that were not, I found the most prevalent reason to be a lack of management support. Simply put, any organization's leader sets the pace for the troops. Commitment from the top is essential, to inspire your team to put in the effort it generally takes to deliver successful software.

I recall working with an organization whose vice presidents of finance and engineering disagreed on the use of technology in the company. Therefore, their IT department was torn between the engineering applications and the administrative and financial applications necessary to run the company. This meant that project managers had difficulty keeping technical people because they quickly became frustrated with their inability to get the resources they needed, such as hardware and systems software, to accomplish their goals. In short, interoffice politics were rendering the computer department impotent; the computer staff was caught in the middle of an executive tug-of-war.

To avoid such situations, your project must have top management support, which will be reflected in adequate budget, resources, and prioritization of your project. Not only are your chances for success greater when you have your management support, but developing the software becomes easier, too, because the team is happier. And a word to the wise: when executives move to other positions or companies, as often happens in this era of mergers and downsizing, check in with the new leaders to be sure that your project's position in their strategy has not slipped in importance.

On occasion, you'll also be faced with the situation where the resources you need to achieve success are controlled by others, who find it politically advantageous to withhold them. This obstacle is not insurmountable, but requires careful planning and patience. You can cleave the project into segments, to reflect what can be accomplished successfully, and eliminate battles over territory.

Let me demonstrate what I mean. Recently, we had the responsibility of implementing a companywide information system. The goal was to provide each department with a set of tools that would enable online entry and retrieval of integrated information, such as project costs, including labor hours expended. The director of finance and the director of operations disagreed over sponsorship of the project (both wanted to claim credit if the project succeeded, though neither would take responsibility if it failed, which looked like a certainty because they couldn't even agree on the day of the week!). The project was stalled. Finally, taking a Solomon-like approach, we cleaved the project into two: a data-gathering and coordination portion, which fell under the responsibility of the director of finance, and a programming project, for which the director of operations was responsible. This allowed the projects to proceed, and both directors could claim success for their portion of the project.

The important issue here is to assess the politics that might affect your success, then plan around them.

Setting Expectations

The users—not you—ultimately determine whether your project is a success, based on how well it meets their expectations. You can help them by setting expectations to realistic—and therefore attainable—levels.

After identifying the people who are stakeholders in the system, it is important to work with them to ensure that they have a shared set of expectations regarding what the software is actually going to do. For example, whereas the executive vice president may be anticipating an Internet-based order system directly into his or her company's parts inventory, the company's purchasing agent may be imagining using an online purchasing system to others' inventories, to save time. Obviously, the question you, the project manager, must answer is: What is the scope of the term "online ordering system?"

A team meeting, or one-on-one meetings with the stakeholders throughout the company, from administrative assistant to executive, can help you get started answering the question. What will come out of the process is a clearer and shared understanding of what the software is intended to do (and not do). A shared vision from the outset makes it easier to achieve success.

At these meetings, it's also a good time to discuss expectations of procedural and systemic changes. When you bring in new software, it affects the processes that surround them. I like to think of programs as the inner core of the total system, circled by concentric rings, as drawn in Figure 2-5. When you change anything in one ring, it affects the rings surrounding it; that is, when you change programs, you generally end up changing the manner in which people and other systems interrelate with the programs, and with one another. Some people will be worried about this, so you need to explain at this point that their tasks will be made easier—and more knowledge-based—which is always regarded positively.

For example, my team implemented a records management system for a police department, to replace a home-grown, FoxPro-based database implementation with a vendor's client/server product. Consequently, the records management staff no longer had to enter officers' reports into the system, or retrieve information for the officers based on a mysterious codification that only the clerks knew. Using the new software, the officers could write up their reports and route them through workflow to sergeants for approval, after which the sergeants routed the reports directly into the system—all without human intervention. In spite of its obvious efficiencies, the clerks were resistant to the system because they feared their jobs would become superfluous. To assuage their concerns, we worked through their new roles with them,

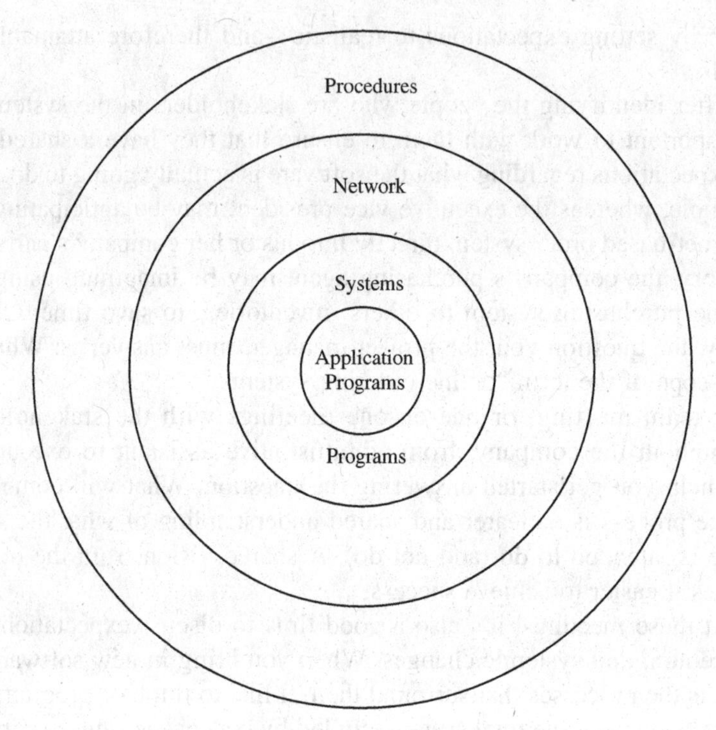

Figure 2-5 *Processes in the total system.*

setting expectations of managing data, assisting with research, and administrating data rather than inputting reports. Ultimately overall, job satisfaction increased. The clerical function of entering data became a task of data content coordination and management.

COMPILING THE PROJECT BIBLE

Every project needs a "bible"; that is, a set of the documents that define and describe the project throughout its life. These documents define why it was conceived, and what it is intended to achieve; the plans for achieving the goals; the parties involved, the specifications, and the schedules; and, finally, the lessons learned, and information for ongoing maintenance after the project is completed.

To compile your project bible, either get out some three-ring binders with dividers, or set up a new file folder on your computer—or both (you'll probably need both if you have any external activities, such as product deliveries and receipts).

Elsewhere in the book I address the specific content of documents; here I want to show you how the documents interrelate. Whichever life cycle definition you use, institute a "tree" for your project along which you'll set down a document for each life cycle phase. A sample is shown in Table 2-1.

A phase can be delineated by:

- A unique audience or set of audiences involved
- Completion of milestone documents or products by the end of that phase
- Separate and distinct uses of the information prepared
- Dependency on preceding phases

Each phase has a distinct mix of reviewers, milestone documents, and purposes, throughout which occur the project management steps—defining, planning, leading, monitoring, and completing.

TABLE 2-1 The Document Tree

Life Cycle Phase	Milestone Document	Review By
Initiation	Project Work Plan, Standards	Program Director, Sponsor, Project Manager, Senior Management
Analysis	User Requirements	Users, Vendors
Analysis	System Architecture,* Product Specifications,** RFP***	Users, Development Management, Management, Marketing
Design	General Systems Design, Application Design, Specifications,** RFP***	Development
Modification	Detailed Design, Database and Processing Specifications	Development
Implementation	Test Specifications, User Documents, Operations Manuals, Training Documents	Quality Assurance Users Maintenance Staff
Review	Lessons Learned, Final Report	Users, Maintenance Staff

*This document is not necessary if you are creating stand-alone software. However, if you are developing software that will run on a network, even a single page describing how the software will use network services will be helpful for others—operations staff, for example.

**These documents are necessary only if you are "productizing" your software—you'll be selling to others as a software product.

***The RFP can be part of the design phase documentation if you are selecting COTS, or during analysis if the vendor is to do the design.

I set up my project bible folders to include all such documents, as well as:

- Sign-offs and authorizations for documents
- Meeting minutes and issues lists
- Vendor evaluations
- Correspondence
- Notes, including records of critical phone calls
- Change orders

More than once, others will refer to the project bible for information as to why design decisions were made, why change orders were authorized, and what the scope was to have been.

CHOOSING PROJECTS WITH THE BEST CHANCE OF SUCCESS

You'll enjoy managing software projects more, and everyone working with you will be more likely to enjoy the process as well, if you begin with a feasible and achievable goal that is coupled with a well-considered work plan to achieve that goal. (Having said that, not every software project is do-able; and others are workable but miss the mark for a variety of reasons.) In this chapter, I'll introduce ways of increasing your chances of working on projects that will succeed.

Requests for Proposals and Proposals

We begin with requests for proposals (RFPs) and proposals themselves. Whether you are managing an internal or external software development effort, the information you need for RFPs and proposals is the same; the only differences will be in terms of legal issues that naturally occur when two organizations are contracting with one another. In an internal proposal, you will still be expected to "spin" benefits of the proposed project, although you won't have to give as much company background as you will when you're involved in an external bidding situation.

You will also have to learn to deal with proposals submitted in response to RFPs that *you* initiate. Understanding how they should be prepared will make you a better evaluator, as well as a better proposal writer.

The proposal itself bridges the definition and planning phases of a project. A project proposal defines the work that is authorized by a work order (internal) or contract (external). The proposal defines the project's cost, schedule, and performance, subject to final negotiation and inclusion in the contract.

One Size Doesn't Fit All

I've said it repeatedly: every software project is unique; it stands to reason then that every proposal will be, too. And, after evaluating hundreds of proposals over the years, I'm still amazed at how different proposals can be even when they're written in response to the *same* RFP!

Furthermore, not all companies respond to RFPs with a proposal, even though it's much easier with contemporary tools to recycle parts of previously submitted proposals. In general, you should submit a proposal only when:

- The project is consistent with your long-term goals.
- You have distinctive expertise in the requested area.
- You have available technical resources to develop the software.
- The budget, if stated, is consistent with your cost structure, and is achievable.
- Your costs to maintain the software, or client relationship, will be consistent with your long-term goals.
- You have a good chance of winning the assignment.

Read the RFP Thoroughly

Well-written RFPs stand out immediately from the unqualified, so it still surprises me how much time and money companies spend on submitting responses to RFPs for which they're not qualified. The motto is: If you are a sow's ear, don't try to pass for a silk purse. We know the difference.

If you are sure you qualify, rate yourself against your competition. How distinctive are your capabilities? If resumes of proposed staff are requested, as is typical these days, do yours match the requested requirements?

When I write an RFP, I ask for relevant experience; firms that are selected for an interview are asked to demonstrate their software. If they are developing software specific to the current environment, ask that they demonstrate something similar. The point is to verify claims of competence whenever it's possible and as early as possible. The risk of finding out *after* you've contracted with someone that he or she isn't all that was promised should be obvious.

When you don your proposer's hat, you'll need to ascertain that the RFP demonstrates a real requirement. Also, find out whether there's a budget for the project. Be aware that sometimes an RFP is written to gather information, not to actually launch a project. In short, it's not real. Frequently, the RFP will reveal enough information so that the company realizes it can take on the project with its own resources. So it's important to examine the RFP carefully, to determine the viability and importance of the proposed project.

Understanding the Proposal Process

Once you determine that you can offer a distinctive capability in response to the RFP, then a proposal might be in order. I say "might," rather than "is" here, because there are other considerations involved.

The following steps were designed to help you decide when to submit proposals; that is, to ensure that you submit proposals for projects you have a good chance of winning. During this process, whenever possible, review your approach and draft documents with the project sponsor. Not only will this reduce the possibility that you'll go off in a nonproductive direction, but you'll generally gain greater clarity on the project at hand, as well as develop more familiarity with your sponsor. Figure 2-6 shows the steps in the process.

1. *Conduct a background check.* Find out whatever you can about the company (or agency, if you're dealing with the public sector). Attend the bidders' conference, if one is convened. If you know people who have worked for the issuing company, find out what is happening (is the company growing, doing well; is it a leader in its field?). Search the Internet for background on the company and for any recent news articles on it. Not only will this give you information as to where the project fits into the company's goals, but it can also help you identify a winning theme for your proposal, such as lowering cost, or time-to-market.

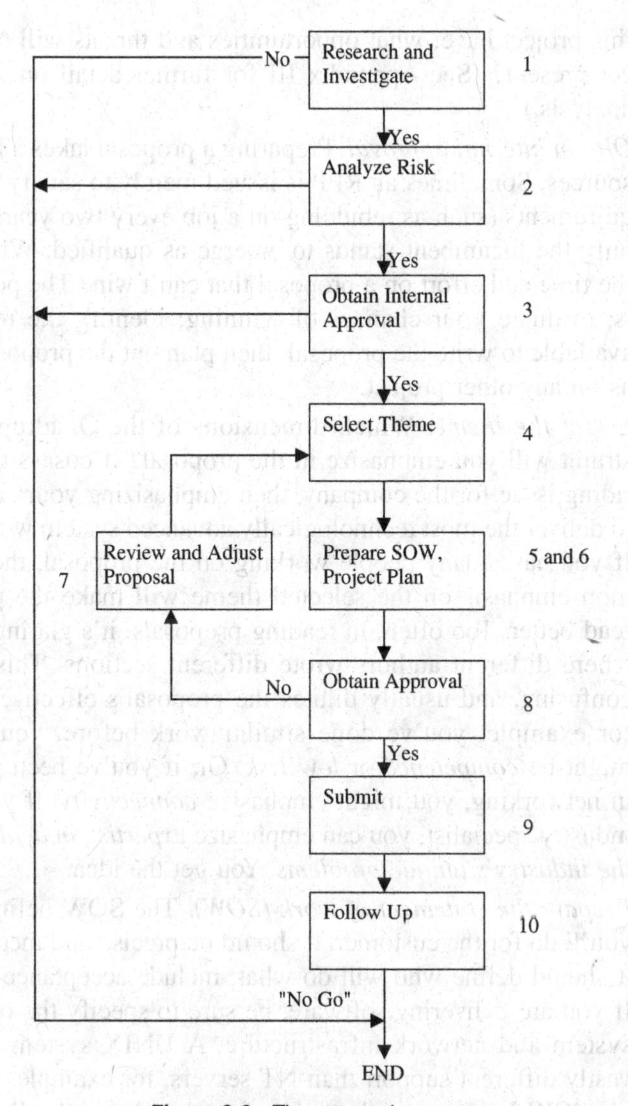

Figure 2-6 *The proposal process.*

2. *Analyze the risk.* Especially in software projects, the risk of failure may outweigh any profit potential. Risk is addressed in great detail in Chapter 3, but for proposal purposes, know that you should go at least to the level of performing a *strengths, weaknesses, opportunities, and threats* (SWOT) analysis prior to going any further. Ask: What strengths and weaknesses does

this project have; what opportunities and threats will this project present? (See Appendix III for further detail on a SWOT analysis.)

3. *Obtain internal approval.* Preparing a proposal takes a lot of resources. Sometimes an RFP is issued merely to satisfy legal requirements (such as rebidding on a job every two years), when only the incumbent stands to emerge as qualified. Why waste the time and effort on a proposal that can't win? The point here is, evaluate your chances of winning; identify the resources available to write the proposal; then plan out the proposal effort as on any other project.
4. *Select the theme.* Which dimensions of the Quadruple Constraint will you emphasize in the proposal? If cost is the overriding issue for the company, then emphasizing your capability to deliver the most technologically advanced system won't help. If you have many people working on the proposal, their common emphasis on the selected theme will make the proposal read better. Too often, in reading proposals, it's glaringly clear where different authors wrote different sections. This can be confusing, and usually dilutes the proposal's effectiveness. If, for example, you've done similar work before, your theme might be *competence* or *low risk.* Or, if you've been pioneers in networking, you might emphasize *connectivity*. If you're an industry specialist, you can emphasize *expertise in dealing with the industry's unique problems*. You get the idea.
5. *Prepare the statement of work (SOW).* The SOW defines what you'll do for the customer. It should be precise and measurable; it should define who will do what; include acceptance criteria. If you are delivering software, be sure to specify the operating system and network infrastructure. A UNIX system requires vastly different support than NT servers, for example. (A sample SOW I used recently can be found in Appendix II.)
6. *Develop a project plan to accommodate the Quadruple Constraint.* The proposal is, as stated before, a bridge between the definition and the planning phase. The SOW defines the project, while the *project plan* says how you'll accomplish it. You'll find a sample of the work plan I use in Appendix I. Note that it contains a schedule; lists deliverables and responsibilities; identifies resources required; and relates to the SOW. It allows the requestor to see how well you understand the RFP.

The planning process is described in greater detail in Chapter 3, but suffice it to say here that you must address your solution to satisfying the Quadruple Constraint in your proposal. First, you address the performance dimension by using the *work breakdown structure* (WBS) to identify the tasks to be performed. These tasks should take no longer than two weeks; they should have quantifiable inputs, outputs, schedules, and assigned responsibilities. For example, the statement "implement system" is too broad a WBS task. Instead, you might divide it into three tasks: install software at client site; test software; accept software. These are distinct tasks that can be monitored more efficiently. You then address the time dimension, by laying out the tasks according to time, using network diagrams. You evaluate the level of risk acceptable for the project type, and include additional tasks and time to accommodate risk-reduction activities. Finally, you cost each activity, to determine the budget constraint. A tool such as Microsoft Project enables you to represent the WBS, the schedule, and the resource usage for your project.

Be sure to include a *project organization chart* in your work plan, not merely a corporate organization chart. While the latter can help clarify roles in a joint venture, most of the people evaluating your proposal will never get an opportunity to meet your company executives, but they will have day-to-day interactions with the project people and organization you propose.

7. *Review and adjust the proposal*. You may find that the project will take too long, and that you'll have to shorten the time allotment for some tasks. Using your project management software tool, you can quickly stage what-if scenarios to determine the best sequencing of tasks. By the way, if, during planning, you determine that you really can't undertake the project, this is as good a point as any to stop. It's also a good point to review the proposal for completeness, consistency, and accuracy. There is never any excuse for misspellings or for failing to insert the correct company name in a proposal template you're reusing. I've seen too many of those!

Often, you'll hold off writing the proposal until the last minute, while awaiting pricing of components, for example. A word of caution: failure to allow for adequate review time may result in major errors—such as misquoting hardware to be delivered, or omitting critical software products, such as additional copies of SQL Server licenses and server operating system software that are also required. Remember, you'll be liable for these errors.

8. *Obtain approval.* Typically, the company head or marketing director will approve the proposal; and someone from upper management usually must sign the eventual contract or agreement if you win the assignment, so he or she is usually involved in the proposal review process to prevent any last-minute problems. To ensure upper management involvement, in our RFPs we include a letter that must be signed by a responsible executive. It's always wise to keep management informed as you proceed, so that they are not surprised later in the project. You can get formal sign-off by submitting a brief that describes the proposal scope, the project's risks and benefits, and the resources to be committed to the project. Text Box 2-1 has a sample of such a brief.
9. *Submit the proposal.* Today, with overnight express services and online tracking capabilities, I'm still amazed when a company spends much time and effort on preparing a proposal, then scrimps on the shipping costs, resulting in a proposal that arrives too late for consideration. From the outset, plan how you'll deliver the proposal, usually in multiple copies, promptly. Confirm with the recipient that it has arrived.
10. *Follow up.* If your proposal does not win, don't forget to follow up with the recipient to find out why you were not selected. It will provide valuable feedback for the next time you go through the proposal process.

If your proposal is favorably received, the next step may be a face-to-face interview, during which you demonstrate your software and systems. In today's communications-friendly environment, you can use a browser through the Internet to effectively display your "softwares," without having to bring in servers, or limit what you can display.

Keep in mind, you can rehearse the contract negotiations. Often, I make up a list of questions I suspect might be asked of us during this process, and determine acceptable answers. A starter list of possible questions can be found in Text Box 2-2.

Of course, your goal should be to rehearse your presentation until it's perfect—hopefully, perfect enough to win you the project. Then, of course, you'll have to negotiate the contract or agreement (contract for products, agreements for services). By involving your senior management in the process early, no surprise clauses in the agreement will preclude authorized signature. (I generally include a sample copy of the agreement in RFPs to prevent such problems.)

Checklists:
As you go through the planning process, you'll identify tasks that need to be performed each time you prepare a proposal. Since it's easy to overlook items when you're hurrying to complete the proposal, you may want to create your own checklist of items that should be addressed and included in the proposal. A typical checklist is given in Text Box 2-3.

Facing Proposal Challenges

The following are common challenges you'll encounter in the proposal process:

- *Balancing your desire to demonstrate your competence in the proposal against doing the entire assignment in the proposal.* This often occurs when you plan the project bottom-up. Plan only to the level at which you demonstrate that you've dealt with problems of this type before. Include a work plan from a previous assignment, for example.
- *Not doing enough planning in the proposal.* A poorly planned proposal will not win you the assignment, and it can damage your reputation. You should plan for any significant issues to the degree at which you are sure you can do the job properly. Otherwise, do not submit the proposal.
- *Having to rush to complete the proposal and submit it on time.* Prior proper planning can eliminate this problem.

TEXT BOX 2-1 PROJECT SUMMARY BRIEFING

Airport Management System Implementation

Action Requested:

Proposal Approval, Authorization Letter Attached

Project Description:

To implement an automated system to manage airport operations reporting and concession leases at ABC Airport. The cost proposed is $250,000.

Project Scope:

This project covers the implementation of software to automate all agreements between the airports and its tenants, including food, automobile rental, and airlines. It also includes the calculation of shared-use charges based upon passenger activity, such as embarkation and debarkation.

The automation of these processes will include the preparation of billing amounts, but the bills will continue to be prepared by the existing airport accounting system. A flat-file interface between the two systems is required until the accounting system is upgraded, at a later date.

These functions are precisely those provided in our packaged software. Training and implementation resources are also requested on a fixed-price basis. User acceptance is the standard 30-day bug-free operation.

The proposal to be submitted responds to the airport's requirements, using our standard Airport Billing Management System (ABMS) software. No enhancements are required. However, the operating platform is Windows NT, and our NT software is still in beta test until after the proposal due date.

(continues)

TEXT BOX 2-1 PROJECT SUMMARY BRIEFING *(continued)*

Benefits and Risks:

The benefits of proposing on such a system are:

- The NT marketplace is increasing, over 10 percent annually for the last two years.
- If we do not propose, competitors will surely gain market share by adding this client to their list.
- We do not have a presence in ABC region, and this will familiarize the airport with our company name, giving us an opportunity to follow up on other possible projects.

The risks if we do win this assignment are:

- Our NT implementation will take longer to develop than promised in the schedule submitted.
- The software will not run as efficiently on NT, and the client will be displeased, possibly resulting in litigation.
- We will not be able to obtain sufficient technical resources to initially service the airport.

Resources:

Our new NT development team has surpassed projected schedules, and consistently come in within budget in transferring the software platform. The number of bugs have been 80 percent less than forecasted, and staff is available to move the NT version out to the client. Additional documentation has been consistently prepared with the project so that additional support personnel will not be required. The R&D funding required to move platforms is at 60 percent utilization.

TEXT BOX 2-2 SAMPLE PRE-INTERVIEW QUESTIONS

Why did you propose the mix of services?
How available will your staff be?
Why should we use your software rather than a competitor's?
How frequently do you release new software?
How do you handle site-specific customizations when you upgrade?
What is your maintenance charge philosophy?
Who owns the product source code?
Can the source code be placed in escrow?
Do you have a users' group?

TEXT BOX 2-3 CHECKLIST FOR PROPOSAL INCLUSION

Letters and forms with proper signatures
Current names, phone numbers, and addresses of references
Accurate number of references
Ownership of software provided
Specific list of services to be provided
Limits to services, especially on fixed-price proposals
Accurate specifications of platforms required
Complete specifications of additional software or hardware required for proper installation
Availability of key staff
Resumes of key staff to address specific client requirements
Acceptability of client agreement formats and terms
Incorporation of any subcontractor requirements, such as DBE and WBE
Cost-of-living increases (for projects over two years)

- *Lacking management support.* Support can be lacking from your own management and from the company for which you're writing the proposal. In the first case, you may not be able to get the resources you need to execute the assignment; in the second, you may not get the assignment because management has another candidate in mind.
- *Treating internal assignments casually.* Internal projects often have highly political implications. A manager's pet project for which no cost-benefit analysis and planning have been done often becomes a political liability when that manager is replaced by someone with different priorities. And if you cannot secure a budget for your project, it is a strong indication that the support you'll need isn't there from the outset.

NEGOTIATING FOR SUCCESS

Today's litigious environment has made it necessary to scrutinize any agreement, or contract, from the start of the negotiation process. In the case of software projects, you should include one in the RFP itself. If there are exceptions to the agreement provided, it's best to address them early. Clearly, if there are unacceptable clauses, such as regarding source code ownership, that cannot be remediated, then don't submit a proposal.

For both internal and external assignments, if you cannot provide the level of service requested, your only choice is either to negotiate the level of service, or performance, down to a feasible level, or simply decide not to propose.

The Best Contract Forms

When dealing with external organizations, a written agreement is required. Always. Without exception. A written agreement unites all involved parties, the specific individuals of which will change over time, in a common perception of what the union is intended to accomplish. On an internal assignment, it is also a good idea to specify service levels, even for an information systems department serving as a resource to the rest of the organization. Doing so helps frame other departmental expectations—specifying what they can expect from you. Never enter into an agreement without documenting what the "goes-intos" and the "comes-out-ofs" are going to be.

The major types of contracts are:

- Fixed price (FP)
- Cost plus fixed fee (CPFF)
- Cost plus incentive fee (CPIF)
- Time and materials (T&M)

FP contracts are better for the customer, because the maximum financial exposure is clear. CPFF and CPIF contracts similarly limit the customer's financial exposure to actual costs, but they include a known fee or a fee dependent upon achievement of a specified goal. T&M contracts, often called blank checks for the contractor, pay for all time and project costs. Because approval of a specific amount often is required, as in public agencies, T&M contracts are often awarded on a "not to exceed" (NTE) basis. When this happens, you should consider an FP contract, as you stand to lose any profit you might have made on the FP basis by bringing in the project with fewer hours and lower cost.

A CPFF contract is better when:

- Project costs are highly variable.
- You have no control over project costs.
- You want to be assured that you lock in a certain amount of profit.

A CPIF contract is preferable when:

- Project costs are highly variable.
- You have no control over project costs.
- You have a high probability of achieving goals valued by the customer, such as bringing in the software faster, and/or more bug-free, and/or at a greatly reduced cost.

Choosing a Contract Type

Regardless of the contract type you use, it will influence the manner in which you allocate resources to your project. But how do you know which contract type to use, and when? Here are some guidelines.

T&M pricing is better when the project involves:

- A first-time implementation of the software, or a major component, such as platform.
- A first-time customer.
- A high-level service component, and the term of the assignment will be more than six months.
- Multiple design components, over which you have little control, such as new communications (wireless technology), or subcontractors determined by the customer, not you.

An FPP contract is better when:

- You've performed this type of development or installation before, preferably with the same staff.
- You've worked with this customer in the past, and know the host environment.
- There is little risk of failure.
- Software is to be phased in, rather than implemented all at once.
- You are involved in the customer design process.

Increasingly, as companies and agencies seek to limit their exposure on software projects, you'll be asked to submit an FP. This is particularly useful when dealing with software that has been implemented before on similar infrastructure. You can establish a per-server and a per-workstation cost, then extend it by the number of such workstations and servers in the company. However, keep in mind that risk increases as the unknowns increase—for example, a new database management system, a new backup system on the servers, a new operating system, moving the software from a UNIX to an NT platform. To accommodate the risk, you can either charge more or take on a portion of the job on a T&M basis. You, of course, must balance your risk in providing the software and service with the customer's objective to avoid paying more for the project than originally specified.

The CPIF structure is similar to CPFF, except that the amount of the fee depends upon some incentive. In both cases, a fee is added to the actual costs you incur. That means that no matter how much the project costs, you will always recoup your costs and not lose money; however, it also means a limit to the amount of profit you'll make.

Typical incentive fee situations are related to bringing in the software ahead of schedule, or to a percentage of dollars saved by using the new software.

The T&M fee structure is least risky for you, but carries the greatest risk for the customer. Therefore, you may find this hard to implement, especially in public-sector software projects, where staff is reluctant to go back to its governing board for authorization to increase budget.

If you are doing research and development (R&D), here too the best contract form to use is T&M, because the performance and schedule dimension in such projects cannot be unambiguously specified. If, however, you can specify the performance dimension, but not the schedule, then the CPFF or CPIF structure will be best. If, on the other hand, you can specify the performance dimension, then FP is best.

We use FP for the analysis phase of our projects. The risk is lower because we are not actually implementing, merely defining, the project and the requirements. Depending upon the results of the analysis phase, we might use CPFF, CPIF or FP. We rarely do any T&M work because clients are not willing to take the risk of cost overruns.

THE VIRTUAL PROJECT TEAM

The virtual project team concept is a twenty-first century phenomenon, already in wide use in the construction, graphics, and other industries. A virtual project team is one whose members are not collocated. The team, in fact, may be scattered throughout the country or world. Using the Internet, the team exchanges information, collaborates, and produces its deliverables as if they worked side by side.

The implications for contract negotiations of working with a virtual team are that you will, in all likelihood, be required to attend face-to-face meetings with the customer at various points during the projects, so you will need to price these meetings in your proposals accordingly, to include the travel expenses. You will also need to allocate time for travel, for your project managers and, at critical points in the project, for key architectural or design personnel to attend, and for trainers to come on-site. These events should be included in your WBS and in your budget. Chapter 4 has more information regarding virtual teams that you might want to organize.

HANDLING THE PRESSURE

The most important point of this section is that you cannot change the performance specification or the schedule without also changing the cost. In fact, you cannot change any one of the Quadruple Constraint dimensions without concomitant changes occurring in the others. Typically, the customer will try to get you to do more for less—and faster! You can survive this pressure by making sure you've established the minimum acceptable position your company will take. This determines how far you can negotiate without rendering the project unworthy.

Unless an increase in scope is agreed to, generally the negotiated price will be less than what you proposed. To expedite this process, have a spreadsheet preconfigured with component costs. This will enable you to determine the bottom-line effects of changes, and to stage what-if scenarios during the actual negotiations.

Throughout the negotiations, beware of conceding services or product unilaterally. For example, let's say you have bid $5,000 per server for software, and you go down to $4,000 during negotiations. The customer may take this move as proof you overbid on *all* components of the proposal. The effect will be exactly the opposite of what you intended. Generally, if you agree to lower a price, you should also alter the performance dimension; for example, reduce the on-site training you proposed, or have customer personnel, rather than yours, train the remainder of the department staff. And be sure to point to potential increases on the risk dimension of the project, if cost and/or schedule are severely decreased.

In general, you will negotiate, even for internal assignments, *after* your proposal has been accepted. The key to success is to ensure that you don't change one of the dimensions of the Quadruple Constraint without making a corresponding change in the other dimensions.

Managing the Legal Details

The agreement you and your customer sign is a legally enforceable document; it defines the financial relationship between both parties, and provides the interpretation of the Quadruple Constraint for the project. Consequently, the agreement is not a place to misrepresent your software's capabilities, your personnel's qualifications, or your financial situation. Doing so can have serious consequences, including the failure of your company. Certainly you can be held responsible for the costs of a replacement should you default, and you may be held responsible for any business losses incurred by your customer as a result of your software failing to perform as specified.

The issues that should be addressed in the agreement will vary according to the type of document, but generally, they should include:

- Ownership of source code
- Ownership of intellectual property and patents, both preexisting and developed during the course of the project
- Licensing rules and restrictions (servers and workstations)
- Effect of using a third-party to update software
- Delivery schedule
- Payment schedule
- Criteria for acceptance
- What constitutes delivery
- Dispute resolution methods
- Security measures, to safeguard software
- Authorized hardware platforms for the software
- Maintenance terms and conditions, and effective date
- Insurance requirements
- Bonding requirements
- Authorized individuals
- Laws of state or country under which the agreement is enforceable
- Termination of agreement
- Nondisclosure of confidential information and use

You may find additional items to include, depending upon the type of project you undertake.

> *Note:*
> The Internet is extending the areas a project can serve. Someone in Tel Aviv can download an update to your software as easily as someone from your home state. Therefore, you'll need to address foreign laws if you are going to serve other countries. Hire a good legal advisor!

Understanding Both Sides

So far, I've presented the proposal process as if you are the proposing company, which often will be the case. But you can—and will—find yourself on the other side of the process, for example, when you issue an RFP for outsourcing software and/or hardware services as part of your own project. In this case, you need a process for evaluating the proposals you'll receive in response to your RFP.

Prior to the receipt of the proposals, meet with your project team to determine the evaluation criteria. These criteria should reflect the important items in your RFP, so that you can later evaluate how well the respondents addressed your requirements. For example, if proximity to your office were a criterion, a respondent who lived within the 25-mile radius you specified would be rated higher than someone whose office was in a foreign country and who had no local representative. A sample evaluation matrix is given in Table 2-2. Typically, 7 to 10 categories are specified in the matrix, depending upon the particular RFP. This initial round of ratings should select out the best of the proposers, those you should interview.

Depending upon the complexity of the task, you may want to allot anywhere from one and a half to six hours for each interview. Generally, the proposer will want to describe his or her company and team, and that alone can take 30 minutes or longer. To make the most of everyone's time, my company decides upon a specific agenda and time limit for each team, with an hour or longer in between each interview. Typically, we have similar, but individual, evaluation matrices for the proposals themselves, and for the subsequent interviews. A sample interview evaluation matrix is shown in Table 2-3. The purpose of the interview matrix is to determine how well the proposer answers questions we may have about the proposal itself, team organization, actual method of performance of the assignment, and software demonstration. We're also interested in the intangibles—the chemistry. As we all know, a paper representation rarely tells the true story of anyone.

For both ratings, with my team, I set down any reasons we found during the evaluation process that might prevent the proposer from receiving a perfect score in a given category. We list these reasons in a separate document, which becomes part of the project documents. Should a vendor ever inquire about his or her failure to be awarded the

TABLE 2-2 Proposal Evaluation Matrix

Factor	Company 1	Company 2	Company 3	Company 4	Company 5
☐ Vendor's Experience and Overall Qualification, including:					
—Financial stability	9	9	1	10	10
—Local presence	8	10	10	4	6
—Experience implementing similar systems	10	10	9	3	10
—References	10	10	9	4	10
—Training and support capability	10	7	6	4	10
☐ Software to be Provided					
—How well the proposed software meets or exceeds RFP requirements	10	9	9	4	8
—Completeness of response	10	8	9	4	10
☐ Services to be Provided					
—Demonstrated understanding of our needs	10	8	8	3	10
—Quality of the overall work plan	7	7	3	5	3
—Completeness of response	9	7	3	3	7
TOTAL	93	85	67	44	84

TABLE 2-3 Interview Evaluation Matrix

Criteria	Weighting (0–5)	Company 1		Company 2		Company 3	
		Score (0–10)	Weighted Score	Score (0–10)	Weighted Score	Score (0–10)	Weighted Score
Functionality of software for the requirements specified	5	9	45	9	45	8	40
Ease of use	4.5	10	45	6	27	6	27
Flexibility	3	8	24	7	21	6	18
Vendor support/training	3.5	9	31.5	7	24.5	6	21
Hardware and systems software	1.5	10	15	10	15	9	13.5
Security/recovery	2.5	10	25	10	25	8	20
Ease of deployment	2.4	8	19.2	7	16.8	6	14.4
Open systems architecture	1.2	9	10.8	10	12	8	9.6
Obsolescence	1.1	10	11	10	11	8	8.8
Total cost	1.2	10	12	8	9.6	6	7.2
Total	25.9	93	238.5	84	206.9	71	179.5

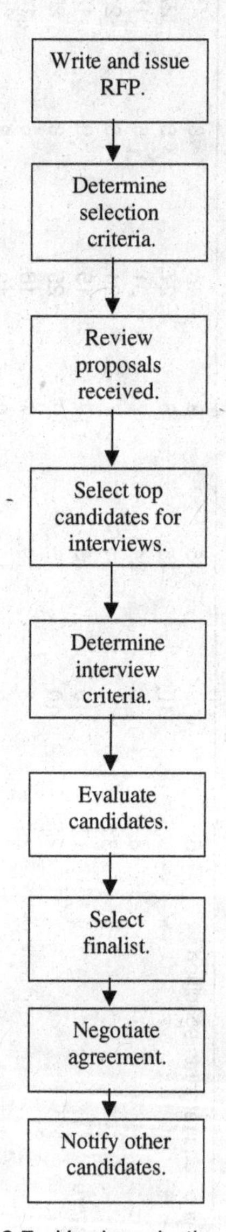

Figure 2-7 *Vendor selection process.*

job, we can refer to that document, which might reveal ratings such as these:

Financial stability. 1—Proposer did not provide any information at all; could have provided something separately in sealed envelope.

Experience. 8—Only one reference was provided, not the requested three.

We also include on the rating document any overall strengths in the proposal, such as:

Overall strengths. Local presence; entire team has worked together before.

After you've made a selection and negotiated your agreement, it's a good idea to send a letter to the proposers who were not chosen. Often a candidate not chosen will wish to be debriefed as to why. I generally review the evaluations with the person, who always finds the information to be of value. Sometimes the company learns, for example, that it is dealing with outdated market information; or is unaware that its platform is no longer current; or that a competitor has additional features that improve its software significantly.

The entire evaluation process is summarized in Figure 2-7. The main point is to be sure that your criteria are well-considered, and reflect the RFP. The proposers put considerable effort into responding to RFPs; the evaluations should be done just as thoughtfully. Quantifying your evaluations helps everyone.

THE IMPORTANCE OF PROJECT DEFINITION

We've come to the end of the definition process. At this stage, you've defined not only what you *can* do (your proposal) in response to what your customer wants (the RFP), but what you *will* do (the contract). The contractual agreement comprises one specification, one schedule, and one budget, with payment milestones, all reflecting an agreed-to amount of risk.

If you've properly planned your project, the contract goals should be achievable; and the relationship formed should be one that benefits all involved: your customer will get the software properly installed when expected, and you will get paid on time and make an anticipated profit.

I can't overstate the importance of the contractual agreement. Paying attention from the outset to the performance promised and agreed to will prevent unpleasant surprises later. Remember, the definition of scope to which you've agreed is the base of your planning, and your subsequent monitoring. Changes in scope are additions to the project, which then need to be defined, planned, and monitored if you are to properly manage them.

Keep in mind, too, that the scope definition can have unforeseen effects. Does the definition cover postimplementation efforts? For example, if you have acquired unique hardware for this project, who owns it at the end of the project? I can remember quite vividly trying to dispose of hardware following a public-sector project for which no agency was authorized to take the now-outdated equipment. Had we defined up front the disposition of the assets of the project, we wouldn't have spent time later resolving the problem.

3

Planning the Project

Probably, by now, you recognize that proper project planning is crucial. Planning simulates the project; it comprises the written description of how the Quadruple Constraint will be satisfied. The *project plan* is actually composed of four subplans, one for each dimension of the Quadruple Constraint:

- The *work breakdown structure* (WBS) defines the performance.
- The *Gantt chart,* or *network diagram,* defines the schedule.
- The financial estimate defines the cost dimension.
- The risk mitigation plan defines the risk dimension.

These may all be combined into a single project plan, or treated individually, depending upon your organization's guidelines. And, in the absence of guidelines, you can do as you wish!

Before you decide which automated planning tool or representational tool to use, you must understand what these tools are designed to represent. In short, no tool can improve the quality of the planning you have put into it. Though it's true that these tools have improved greatly in recent years, and there is no need to send out charts to be professionally printed when technology can do it more quickly, the thinking behind a good plan still must be done the old-fashioned way. Hopefully, that's what this primer will help you to do!

GETTING THERE FROM HERE

A plan defines how we'll get from where we are to where we want to be. In our case, it will be to deploy packaged software throughout a corporation, to design a new Internet site, or to replace the hardware and network infrastructure with the latest Pentium machines and Microsoft Office 2000. And these objectives are measurable: number of hours of training; number of hours to configure each machine; and how proficient employees must be to use the software, as determined by a questionnaire administered at various points before, during, and after training.

Without a plan, how will we know when we've achieved our goals? Without a plan, how can we check whether we're even on the right path to achieve those goals? The plan is the road map we consult to progress—though occasionally we'll have to take detours (not of our own making) before we reach our destination. Continuing the analogy, we'll have many different roads to choose from, but if we've planned correctly, all roads lead to Rome, as the saying goes. And if we've identified our Quadruple Constraint properly, the best road will be easily identifiable.

For example, my firm recently implemented an airport management system. On this project, the schedule was the most important dimension because the manual billing calculations were causing the airport to lose money on its concessions. The sooner the new software could be implemented, the more quickly greater revenue opportunities could be identified and capitalized on. After the project work plan was agreed to, implementation started with great enthusiasm among the staff who were going to benefit from the system. The enthusiasm was so great, in fact, that staff started exploring the possibility of implementing an online real-time integration with the existing accounting system.

Had we not been guided by the project work plan, which stated that interfacing data by manual handoff with the accounting system was the goal for this project, and that integration was a separate project to be embarked upon later, we might never have accomplished our original goal. Certainly, we could not have done so while respecting the schedule constraint—and our project might not have been judged a success.

The point is, we must focus on our original goal, and navigate toward that goal. If we keep changing course along the way, we might never achieve it.

The Best-Laid Plans . . .

. . . will often go astray. Working on software projects in particular, you will encounter unforeseen obstacles. For example, just as you are ready to implement a new billing package, the vendor announces an operating system software change, to run on Windows NT and Windows 2000 instead of UNIX; they will support the UNIX-based software only for another year. What can you do and what should you do?

This is where the project plan can help. While it cannot eliminate problems that arise, as they inevitably will, you can refer back to the plan for guidance; to see how you prioritized the constraint dimensions. In the UNIX case, you can determine from the plan which is more important, cost or schedule. If, say, you have to get the software in as soon as possible to satisfy some business reason, then the UNIX system, even though it will require replacement in another year, will enable the fastest implementation. It will also give better performance, as changing to the NT version will require working out the bugs inherent in changing platforms. If, on the other hand, budget or risk is most important, you might want to stop all further work on the project, and resume it when the vendor has completed the NT migration and the product has been satisfactorily deployed elsewhere. You might also want to reevaluate other products and open up the proposal process again to other vendors.

RISKY BUSINESS

This is as good a time as any to address the fourth dimension of the planning process: risk, and the evaluation of the risk, inherent in any software project. (Remember, because risk can be affected by one or more of the three dimensions in the construct, I also support the Triple Constraint approach.) Risk is defined as "any threat to the achievement of one or more of the cardinal aims of the project."[1] Software is inherently risky: it reflects an implementation of a vision that has been interpreted by many parties over time, and is subject to change due to advances in technology, alterations in the original

[1] *Managing Software Quality and Business Risk,* Martyn Ould, (John Wiley & Sons, Ltd., England, 1999), p. 42.

vision, or, simply, misunderstandings. Because today's software must run on multiple platforms—hardware, communications, operating systems—and interface with a variety of other types of software, there are many points of potential failure.

How can you, the project manager, incorporate risk mitigation techniques early in a project, and maintain risk awareness throughout? And how can you avoid becoming so risk-averse that analysis paralysis sets in, and you fail to reap the benefits of advances? Eliminating uncertainty means repeating what has worked well before. The software industry, like those of the hardware and networks upon which it runs, is constantly innovating and enhancing its capabilities. Unfortunately, this means that upgrading a version of just one software package is a situation that can be fraught with peril. As is often said in the industry, being on the leading edge of technology generally means being on the bleeding edge—a decidedly uncomfortable place to be! Clearly, to be safely on the leading edge, you must anticipate longer implementation cycles, as you discover hidden "features" of integrated components.

To reduce risk, consider doing one or more of the following:

- *Buy software rather than build it.* You may have to change internal procedures, and integrate the purchased software with other systems already in-house, but you'll reap the benefit of hundreds of thousands of dollars already spent on development of the package. You'll also have on hand a larger number of people who know how to use the software, and who can help you.
- *If you must build software, prototype it, rather than develop it using the Waterfall method.* This involves the users early, and keeps them involved; and if you use an incremental prototyping approach, people can actually use the software earlier in real situations. That means earlier benefits for the company as well, and reduces the risk that the software developed won't satisfy the users when it's finally completed.
- *Set the project schedule as short as possible.* Especially where management changes frequently, shorter-term projects mean a greater likelihood that your sponsor will not change, resulting in your project being cancelled or put on the back burner.
- *If you require a critical resource, such as a subject matter expert (SME), try to condense the resource use into short time periods.*

Each time you need the SME, you take the chance that he or she will not be available, thereby introducing time delays into your project.

- *If you are new at project management, slip the schedule from your original estimates.* As a "newbie," you probably are not aware of the many stumbling blocks facing you, so you need to accommodate your own learning curve. For example, even though you decide to buy a software package, you discover that the help package has not been completed for this version, and that it will not run with existing versions of software found elsewhere on the system.
- *Even if you buy a software package, take into account what that will entail.* You will need to train, document procedures, test, revise, administer contracts, and manage the project—and the external vendors.
- *Test the software, in particular when you implement changes or develop it from scratch.* A new project manager should add more time for testing, to deal with unanticipated test results.
- *If the software or hardware components have not been integrated before, allow more time for testing and integration.* Try to integrate a few components at a time, so that you can more rapidly identify where the problem came from. First-time integration is always riskier.
- *If your team is inexperienced, slip schedule and increase cost.* Again, introduce additional tests to detect problems as early as possible, and to isolate their origin.
- *Even if you are not 100 percent sure you'll need a critical resource, ask for it anyhow.* If you wait too long, the odds are the resource will not be available, and you'll have to slip your schedule.

DEFINING THE PROJECT WORK PLAN IN DETAIL

I've alluded to it, and addressed it directly, though briefly, in Chapter 2; here, finally, I discuss the project work plan in detail. (A sample project work plan, using Formula-IT can be found in Appendix I.) I recommend keeping the plan and any updates to it within a single binder. However, depending upon project complexity and duration, it may grow to include many binders. A typical work plan table of contents is shown in Figure 3-1.

1. INTRODUCTION
2. PROJECT SCOPE
3. PROJECT BUDGET & SCHEDULE
4. PROJECT ORGANIZATION & STAFFING
    4.1. Project Organization
    4.2. Project Staffing
5. MANAGEMENT CONTROL
6. PROJECT MANAGEMENT
    6.1. Project Planning
    6.2. Tracking and Monitoring
    6.3. Progress Reporting
    6.4. Software Support
    6.5. Document Management
7. CONCLUSION
APPENDIX I - JOB DESCRIPTIONS
APPENDIX II - CHANGE REQUEST FORM
OTHER ATTACHMENTS:
    STYLE GUIDE FOR WRITTEN REPORTS
    PROJECT SCHEDULE
    TASK SHEETS

Figure 3-1 *Project Work Plan Table of Contents.*

The project plan covers the entire life cycle of your project, regardless of how many phases your particular life cycle may have and what you name the phases. The project plan expands the Quadruple Constraint; it does not restate the design and other document deliverables produced as part of the project work plan, but defines their content, review processes, and scheduled delivery dates. The plan also defines the roles that parties related to the project are to serve, although it does not identify them by name. The plan defines the change order process, as well as the manner in which the project will be managed.

Task Sheets

Most important to the project plan are the task sheets (a sample is shown in Text Box 3-1). The task sheets are intended to describe unambiguously each of the tasks to be carried out during the plan. Each task is detailed according to these categories:

- Task description
- Purpose
- Scope
- Deliverable
- Content of deliverable
- Work method
- Completion criteria

> *Note:*
> Time scales and other schedule-related information are presented elsewhere in the project plan, so the precise end date of the task is not listed in this part. That way you need not rewrite each task sheet when the schedule slips!

Task sheets apply to the planned length of the project, and are a result of the planning effort required for that phase. Where you have multiple phases, the specifics of which are dependent upon the results of earlier phases, you will not be able to describe in detail the tasks required. You can, however, identify those subsequent tasks in successive updates of the tasks in the project plan when you provide the updated schedule and, if necessary, updated budget.

A list of tasks for our contract management system is given in Text Box 3-2. Of course, you may include any additional ones that might make sense in your working environment.

An important point here is that you should define the tasks *before* you embark upon the project, but not before you know the outcome

TEXT BOX 3-1 SAMPLE TASK SHEET
TASK SHEET (TASK 1)

1. TASK DESCRIPTION

Task 1 is to prepare a detailed project plan for the Contract Management Project. This task is part of stage 1, Project Initiation.

2. PURPOSE

The purpose of this task is to develop a plan defining tasks, timescales, responsibilities, and deliverables for use during the implementation of the Contract Management Project at the company. The purpose of the plan is to provide a clear definition to all parties of what is to be produced, by whom and by when, thereby reducing risk to the project.

3. SCOPE

The project plan will cover:

- Tasks
- Timescale
- Responsibilities
- Deliverables

4. DELIVERABLE

A report defining the project plan

5. CONTENT OF DELIVERABLE

The report will be of the format defined in the company's *Standards for the Implementation of New Systems,* reproduced as follows:

1. INTRODUCTION
   An introduction to the project plan.
2. PROJECT SCOPE
   A definition of the scope of the project.
3. PROJECT ORGANIZATION
   A description of the organization of the project, including the roles and responsibilities of each of the project participants.

(continues)

TEXT BOX 3-1 SAMPLE TASK SHEET *(continued)*

4. PROJECT CONTROL
   A description of the approval process for project deliverables. The description will include the role and terms of reference of each committee and executive authority.
5. PROJECT MANAGEMENT
   A description of the mechanism to be used for managing the project team and for tracking and monitoring progress. The description will include the use of project plans, task sheets, and timesheets.

APPENDIX I: JOB DESCRIPTIONS

Job descriptions for all participants in the project.

APPENDIX II: PROJECT PLAN

The project schedule for the Contract Management Project.

APPENDIX III: TASK SHEETS

A description of each task to be carried out during the Contract Management Project.

6. WORK METHOD

The Project Manager will develop a project plan based upon:

- Experience of previous implementation projects
- The work plan agreed to by the Contract Management team

7. COMPLETION

On approval by the company Program Manager.

of prior stages of the project. The goal is to provide a management guideline, not make busywork for yourself. As you will quickly find out if you attempt to define the tasks before you have enough information from early phases, you will find yourself rewriting your tasks rather than doing the actual work. For example, say you're

TEXT BOX 3-2 CANDIDATE TASK SHEETS

1. Initiation
   - Develop Project Plan, including:
     —Resource Requirements
     —Job Descriptions
   - Define project standards.
2. Analysis
   - Analyze requirements.
3. Design
   - Define data.
   - Define processing.
   - Define technology.
   - Define user interface.
   - Compile system design.
4. Software Selection
   - Identify list of potential vendors.
   - Develop Request for Proposal.
   - Evaluate responses.
   - Conduct demonstrations.
   - Select vendor.
5. Software Development/Modification
   - Prepare for development.
   - Build/modify database.
   - Build/modify processing.
6. Implementation
   - Plan system introduction.
   - Assemble technology products.
   - Conduct user training.
   - Conduct testing.
   - Migrate data.
   - Run parallel/pilot.
   - Initiate operations.
7. Review
   - Review system.

going to design and develop the software yourself, and you define your tasks accordingly. If you then find a package you want to implement instead, your work plan becomes more targeted to managing vendor interface, rather than to your own development efforts.

> *Note:*
> I tend to keep the budget information—along with pay authorizations, encumbrance reports, and invoices—in a separate binder, even though they are part of the work plan. I do this because I do not want financial information to be as widely disseminated as the personnel responsibilities and project tasks. You can however, make this information part of the project plan.

Who Does What, When

One of the most important purposes of the task sheets is to identify who will be doing what, and when in the project. When these decisions are made early enough in the project, users can address such items as testing criteria to ensure that they prepare their requirements with testability in mind. Recently, my firm undertook a project for which we had selected a vendor, who prepared a one-page work plan. While it addressed the major tasks in a simple bar chart format, it didn't specify when the payments would be made, nor what the criteria for payment were going to be. Working together, we arrived at a work plan that clearly identified what the payment criteria would be, using the task concept. (A sample of this is included in Appendix II, to show that there are many ways of representing the information required to ensure smooth execution of your plan.)

Keep Them Straight

Remember, there are separate plans for the performance (the statement of work and the design specification), the schedule (your bar chart or other representation), the risk mitigation plan, and the costs (a committed budget or cash flow) that amplify the project work plan in greater detail. This discussion, however, addresses the project plan that ties them all together, as shown in Figure 3-2.

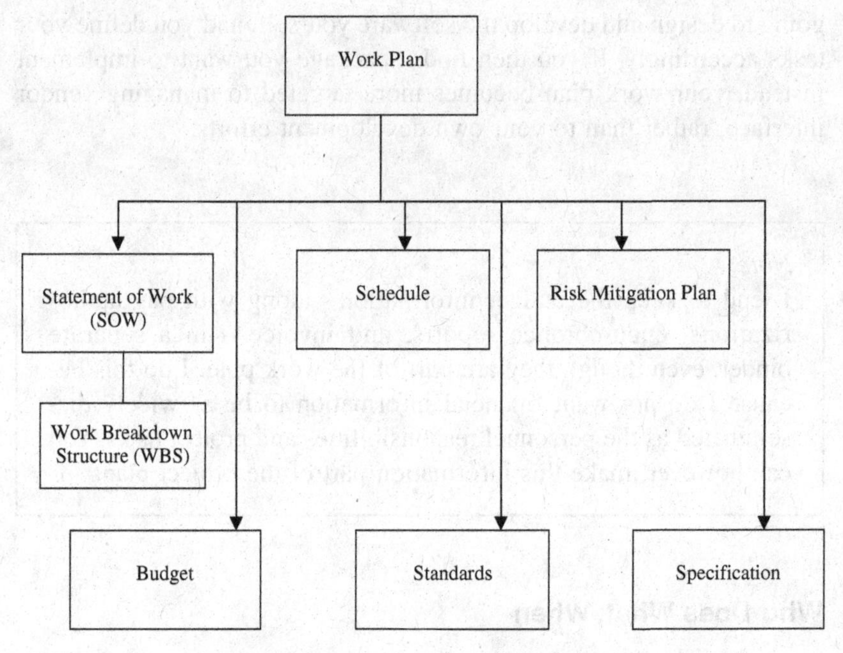

***Figure 3-2** Project planning documents.*

Keep Them Simple

One danger in writing plans is going too far. Overplanning can be as much a waste of resources as underplanning. Think of it this way: spend no more time in planning than you would have to spend correcting the problems that probably would arise from not having a plan. Of course, if you're doing this for the first time, you have no way of knowing when you're overplanning. In this case, it's best to overplan than underplan. A good guideline is that your plan shouldn't be thicker than the software specification.

Share Them

When you've written your project plan, be sure to share it with the people on your team. The users, the sponsors, and the developers will all want to know how the work will proceed, and, more importantly, what their roles are and what they can expect of others. The project work plan serves to set the expectations of all the parties involved in your project. It also allows feedback before the project gets underway, alerting you to any roles or expectations that are unfeasible or unclear.

Statement of Work and Work Plan

As explained earlier, the statement of work, the SOW, is that portion of the agreement or contract that explicitly defines what the project organization will do for, and deliver to, the customer or user.[2] The SOW, typically, is the basis for the task definitions and your objectives—that is, what you intend to accomplish through application of the work plan. In a project internal to your own company, you might include this same information in a memo or work order rather than in a contract. And, note, a complete work plan requires a schedule and budget.

The Project Management Institute (PMI) defines the project plan as: "A formal, approved document used to guide both project execution and project control. The primary uses of the project plan are to document planning assumptions and decisions, to facilitate communication among stakeholders, and to document approved scope, cost, and schedule baselines."[3] PMI defines the SOW as "a narrative description of products or services to be supplied under contract."[4]

Work Breakdown Structure

This section explains how to determine what is a manageable piece of work. The concept of the work breakdown structure, the WBS, is to break down the larger task specified in the SOW into smaller work packages, each of which is understandable, achievable, and measurable. If you or someone on the team can't understand it, you probably need to break it down into even smaller work packages, arriving at a level where it's clear to everyone what has to be done. The purpose is to be sure that you've identified all the required activities, and related them logically, thus reducing the chances that you'll have unpleasant project surprises due to a failure to plan adequately.

For example, when my firm was installing the airport management system I described earlier, we had many different activities requiring completion before the system could be implemented. These included testing and installing hardware, as well as obtaining and testing network services. Text Box 3-3 shows the high-level tasks to be accomplished, and Text Box 3-4 shows one of those tasks broken down into

[2]Lewin and Rosenau, *Software Project Management: Step by Step,* 2nd edition (Marsha D. Lewin Associates, Inc., Los Angeles, CA), 1988, pp. 76–77.
[3]*A Guide to the Project Management Body of Knowledge* (PMBOK® Guide), 2000 edition, Newtown Square, PA, p. 205.
[4]*Ibid.,* p. 208.

TEXT BOX 3-3 HIGH-LEVEL TASKS TO BE ACCOMPLISHED

1. Define airport system requirements.
2. Acquire software.
3. Install software.
4. Train key users.
5. Complete data conversion.
6. Conduct pilot test.
7. Implement airport system.
8. Close out project.
9. Manage project.

greater detail. You can continue breaking down, or decomposing, the functions within each task until you reach a level that everyone understands and believes they can accomplish. For example, the task "evaluate proposals" could be broken down further into:

1. Determine top three.
2. Notify top three for interviews.
3. Evaluate interview finalists.
4. Determine vendor.

Breaking down one of these major tasks into its component parts revealed to us how the services of others not generally involved in software systems implementation would be necessary. Not only did we realize we required an outside contractor to install hardware computing components in the police vehicles, but we learned we also needed to coordinate with the organization's mechanics to allocate time and labor, as well as bays in their shops, to make the necessary modifications on and installations to the vehicles.

How Far Is Far Enough?

As you work with the WBS, you'll discover there is no one magic formula: too many levels becomes hard to manage efficiently, while an overly broad WBS accomplishes nothing in terms of allowing you to monitor your project—you only know when you get to the end, if you

TEXT BOX 3-4 WORK BREAKDOWN STRUCTURE

2. Acquire Software

2.1 Prepare system flowcharts.

2.1.1 Document current practices.

2.1.2 Modify current practices.

2.1.3 Receive concurrence from department.

2.2 Select software vendor.

2.2.1 Circulate RFP.

2.2.2 Evaluate proposals.

2.2.3 Interview finalists.

2.2.4 Select vendor.

2.2.5 Execute contract.

2.3 Receive software.

ever do! The rule of thumb I've used over the years is to tie a WBS task to an individual or to a team, to a payment, or to a schedule completion. For example, if a database server is required to deploy the application software, I would treat those as two individual tasks within the WBS. The server will typically be configured and tested by a group that is separate from the application developers or vendors. And because the vendor will want to get paid for his or her effort in deploying the software—regardless of the failure of other activities within his or her control—by identifying such activities separately, I can better determine if he or she is entitled to payment.

Completion typically is scheduled at the end of a life cycle phase (when you're using the Waterfall model of development, for example), so the task generally does not cross life-cycle phases. Thus, your WBS might start with the project phase, with each major task broken down into smaller tasks. And you would define deliverables at the end of each phase: a document, test results, and a specification. Just make sure you have a deliverable. If you don't have something that can be measured, how do you know whether you're where you planned to be?

Another good rule of thumb to follow is to have a deliverable to mark the move from one project phase to another. This deliverable, typically written, documents the results of the just-completed phase. This permits you to move along in your development methodology with a record of the successful completion of the previous phases.

In software development, these records of accomplishment are very important; as information being handed off to others, who must act upon them, they identify any ambiguities, incompletions, or inaccuracies that must be addressed. This is particularly important in software development because the later a defect is found, the higher the cost of correcting it. Costs can, for example, escalate as much as 10 times higher after coding, and 100 times higher to correct a production error.[5] (Sample end-of-phase deliverables were listed previously, in Table 2-1.)

If you conduct project reviews—typically at the end of a life-cycle phase—you might want to identify them as tasks in your WBS as well.

Summarizing, then, your WBS work "atom" should be guided by the following:

- Any system development methodology is incorporated into your WBS.
- Tasks do not cross project phases.
- Payment milestones can be determined from the tasks.
- Separate management is reflected in individual tasks.
- Document deliverables are milestone tasks.
- Significant project reviews are tasks.
- Completion of a project phase is a task.
- Too many tasks spoil the monitoring.

The WBS is very critical to your project's success, so it's a good idea to have a colleague look it over or draw up an alternative to compare against. This is especially true if the concept of using a WBS is new to you, or you're new to managing software projects. This added input will prevent many mistakes. Just remember to return the favor!

You should also plan for a variety of tests, to demonstrate how well your software has been developed, and how well the components integrate. The amount of testing you do will be based upon the currency of the platforms and programs, as well as the expertise of your development staff. If you've outsourced the development, you'll probably

[5]William E. Perry, *Effective Methods for Software Testing,* 2nd edition (John Wiley & Sons, Inc., New York, 2000, pp. 39–40).

want to spend more time to test the users' procedures as well as the programs.

Finally, it's wise to make a checklist of the tasks you identify in your projects. After a while, you will have compiled a pretty complete set of tasks. Ultimately, you should be able to import your current WBS into your new project's schedule. That said, beware of copying, without appropriate adaptation, one WBS to another project. Never forget, each project is unique. Use your prior WBS task list as a guideline, to be sure nothing falls between the cracks; but you still must accommodate the performance and resource issues that make the new project unique.

SETTING STANDARDS

For those of you who have developed software yourself, I don't have to make a case for ensuring that you have some set of standards against which your software must measure. You know what it's like to spend hours, if not weeks, trying to determine interface (handoff) formats for databases. I confess that when I started programming, I used geographical terms for tag names (yes, countries, with cities, streets, even rivers for levels of indentation). I soon repented after suffering through programs I took over, where someone had tagged paragraphs and phrases using random numbers. I quickly realized that not everyone had the same interest in geography that I did.

So I became a firm believer in standards. In some environments, standards are called guidelines, because they aren't enforceable insofar as a software development Nazi will come and eliminate your code. But as a considerate, professional humane activity, implementing software according to shared standards will make your job and the job of those who follow you to maintain, interface, or integrate your software much easier.

Standards should cover not only the coding and design of the software, but the network and operating systems environment in which the software is to perform. A sample set of standard topics that my firm includes as part of Formula-IT is shown in Text Box 3-5. For highly specialized projects, such as a geographic information system, or for Internet publishing, we add sections to the basic standards document. Each supplement extends the basic set of standards only to the particular subsystem in question. But it saves rewriting the standards for each project!

TEXT BOX 3-5 IMPLEMENTATION STANDARDS

1. INTRODUCTION . . . .
   1.1. Purpose and Scope . . . .
   1.2. The Benefits of Adopting Standards . . . .
   1.3. Structure of This Document . . . .
2. DEFINITION, APPLICABILITY, AND RESPONSIBILITY .
3. STRATEGY IMPLEMENTATION MANAGEMENT AND CONTROL . . . .
   3.1. Project Planning . . . .
   3.2. Tracking and Monitoring . . . .
   3.3. Progress Reporting . . . .
4. COMPUTER OPERATIONS . . . .
   4.1. Computer Operations Procedures . . . .
   4.2. System Acceptance Testing . . . .
   4.3. Logical Security of Applications . . . .
   4.4. Physical Security . . . .
   4.5. Backup and Restore . . . .
   4.6. Archiving . . . .
   4.7. Disaster Recovery Planning . . . .
5. APPLICATION SOFTWARE DEVELOPMENT . . . . . .
   5.1. Specification . . . .
   5.2. Construction . . . .
   5.3. Implementation . . . .
   5.4. Review . . . .
6. NETWORK ADMINISTRATION AND END-USER SUPPORT . . . .
   6.1. Network Security . . . .
   6.2. System Management. . . . .
   6.3. Service Levels . . . .
   6.4. User Support . . . .
7. APPLICATION SOFTWARE STANDARDS . . . .
   7.1. Application Software Selection . . . .
   7.2. Application Software Acquisition . . . .
   7.3. Application Software Deployment . . . .
8. TECHNICAL STANDARDS . . . .
   8.1. Naming Standards . . . .
   8.2. File System Hierarchies . . . .
   8.3. System Availability . . . .

(continues)

TEXT BOX 3-5 IMPLEMENTATION STANDARDS *(continued)*

Appendices

SCHEDULING

The second part of planning is charting the events and activities included in your WBS along a timeline—you schedule them. The secret is to arrange the planned events so that they occur in a logical relationship to one another. For example, you wouldn't go live with your new records management system until you tested it, and before the user has accepted the system. This implies a natural order of events that is a critical part of planning.

There are three approaches to scheduling, each with its own form of representation: bar charts, milestones, and network diagrams. All are supported by automated systems such as MS Project. That said, I find that bar charts are the most commonly used, probably because they are the simplest for people to understand. And simplicity is invaluable when you are attempting to bring diverse groups together, such as in most software projects. After reading the descriptions of all three, you can decide which will work best for you.

Bar Charts

Bar charts (also often called Gantt charts, after H. L. Gantt, the industrial engineer who popularized them during World War I) are frequently used for scheduling. You can create one using any popular project scheduling system; or you can even use Excel to create one for a very simple project. A sample bar chart, representing a plan for implementation of multiple software projects, is shown in Figure 3-3.

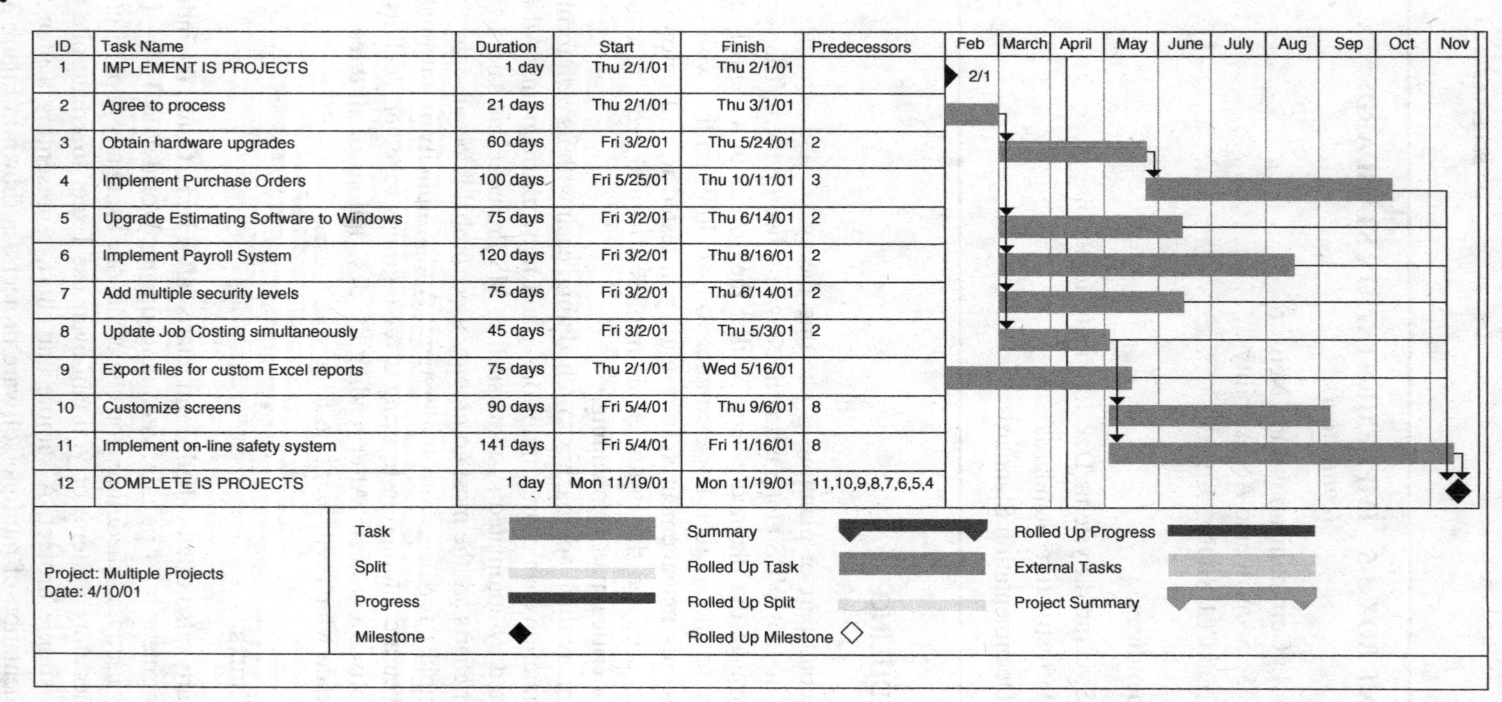

Figure 3-3 *Bar chart.*

Unfortunately, in this book, you can't see the colors available with today's technology. However, even in this black-and-white presentation, you'll note:

- Each activity has its own line, or bar.
- Today's date is clearly marked.
- Performance against planned schedule can be quickly determined for each task.

The problem with bar charts is that, while they can be easily understood by people of diverse backgrounds, they don't really tell you all you need to know. For one, you cannot infer the overall project status, because there's no interdependency of tasks shown. For example, if the hardware being delivered is task 3 and everything else is in place but the hardware, the project is far from complete. Task 4 cannot be completed if task 3 is not. Even though everything else might be complete on a task level, you don't know how your project really is doing.

Milestones

Another way of representing your project is to define its milestones. This you do by to laying them out along the schedule timeline in the bar chart format. A milestone is a significant event in the life of a project, typically the achievement of an activity (such as testing), or the completion and submission of a deliverable (such as the design specification). You will know the major milestones of your project because you, hopefully, will have defined them, early on, in the contract or statement of work. And, as I said earlier, for each phase in the life cycle you've selected, there is one or more deliverable that you can identify as a milestone. A milestone schedule is shown in Figure 3-4.

Milestone schedules are helpful on a project or program (multiple projects that are related) level, but they still do not show you the interdependencies between the tasks and activities. So you will still need another tool to help you determine where your project really stands.

Network Diagrams

And thus we enter the world of network diagrams, which show the precedent conditions and the sequential constraints for each activity. Only with these network diagrams can we determine how the tasks in our project are truly related. This topic is worthy of a book-length

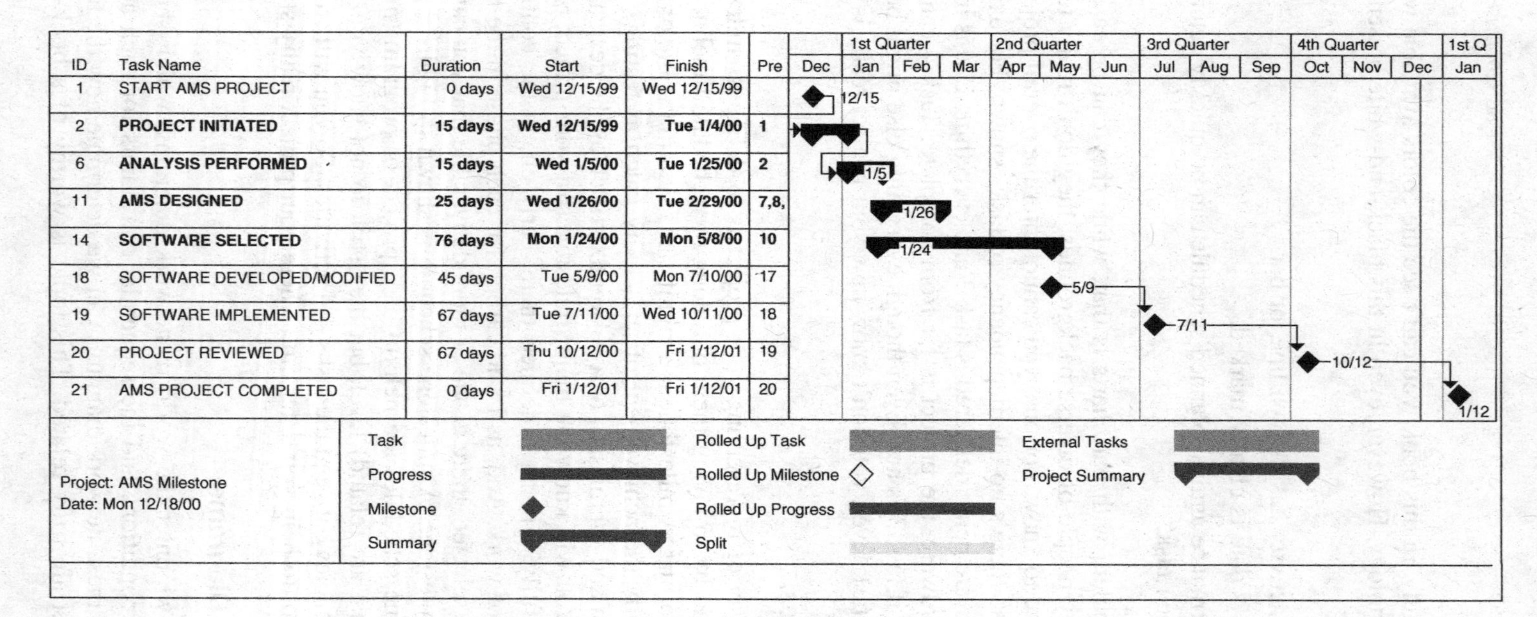

Figure 3-4 *Milestone schedule.*

explanation, so I refer you to the excellent treatment of the differences, jargon, and details in the *Guide to the Project Management Body of Knowledge,*[6] and, especially, Mickey Rosenau's *Successful Project Management.*[7] In my own experience, given the level of today's project management software, you should focus on what works best for your particular project team members.

Network Diagram Types

There are many forms of network diagrams, the most common being Program Evaluation and Review Technique (PERT) and the Precedence and Arrow Diagramming Methods (PDM and ADM, respectively). A few words will clarify how these important event or activity approaches have been merged in today's easy-to-use charting software.

ADM is activity-oriented, with the arrow representing the activity. PDM represents the node (a circle or box) as the actual activity, and links the activity nodes together in a precedence relationship. PERT is a hybrid. See Figure 3-5 for an example of how the representations differ.

I want to stress here that diagramming events by themselves isn't enough; nor is identifying the length of an activity. However, linking what needs to be done with the order in which the tasks should be handled will give you a powerful tool with which you can monitor and effectively control your projects.

In sum, I'd say that, using today's software to achieve the best project management practices, the diagram that will tell you the most the quickest is the time-based precedence bar chart, with activities on the arrows and the milestones on the nodes.

Diagram Comparisons

A picture, as they say, is worth a thousand words; trying to explain the advantages and differences of these diagram types would take even more without an example. So, for clarity, three representations of an implementation schedule for my firm's airport system can be found in Figures 3-6 through 3-8. The first is a simple bar chart; the second is a PERT diagram; and the third is a CPM diagram. CPM (Critical Path

[6]*A Guide to the Project Management Body of Knowledge* (PMBOK® Guide), 2000 edition (Project Management Institute, Newtown Square, PA, 2000.

[7]Milton D. Rosenau, Jr, *Successful Project Management*, 3rd edition, John Wiley & Sons, Inc, New York, 1998.

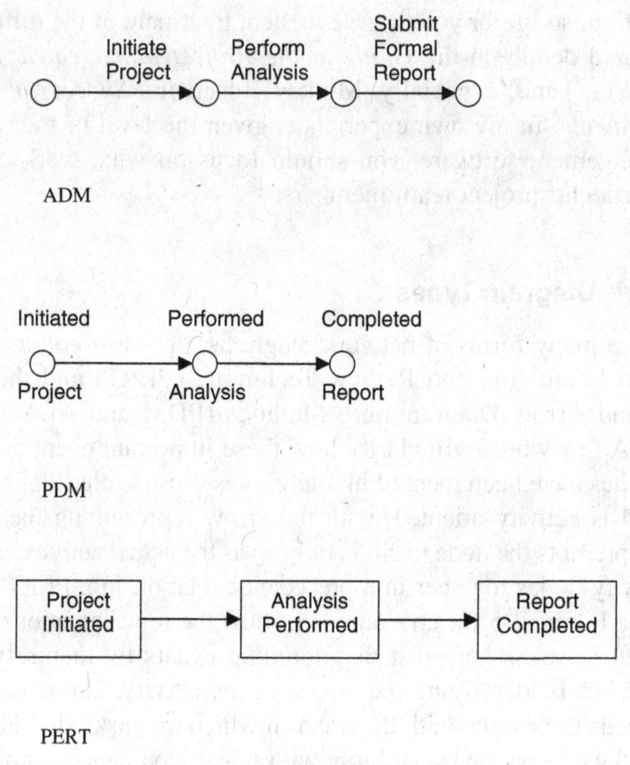

Figure 3-5 *Network diagramming conventions.*

Method) takes the activities and arranges them with precedence such that you can determine where in your project schedule you have the least amount of flexibility. (Now you know what people mean when they say that their task is "critical path.") What's exciting about today's technology is that you can represent the critical path by pressing a single button to elect that option (as opposed to recharting manually!).

Returning to your WBS, you take the list of major tasks, and schedule them. Note in the example bar chart that I've given the tasks an initial estimate of how long each will take. I work in the following manner: I take the date on which I want the entire project to end, usually dictated by contract language or a business requirement for availability of the system, and then work backward. I know, for example, that we cannot test the system until it has been installed; that we cannot validate the data until it has been entered. In some cases, when I lay out the events, such as tasks 7 through 10 in Figure 3-9, the events

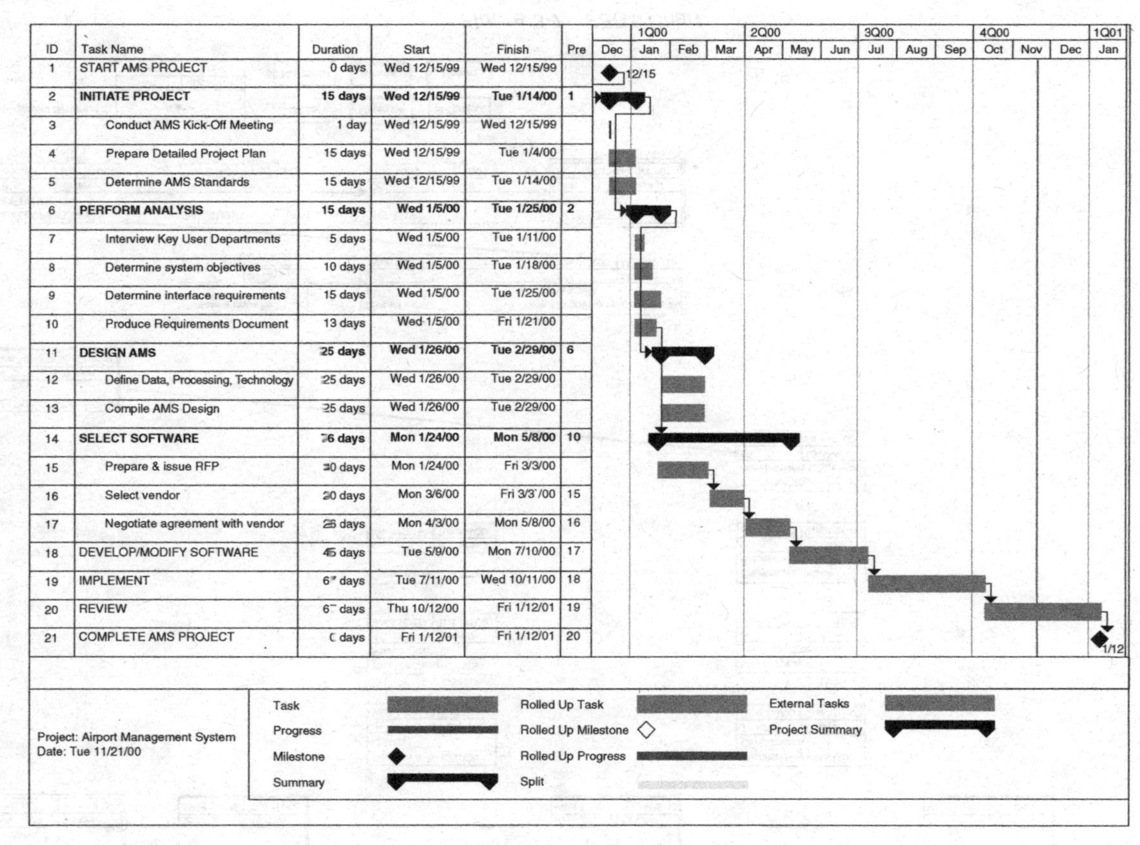

Figure 3-6 Gantt Chart.

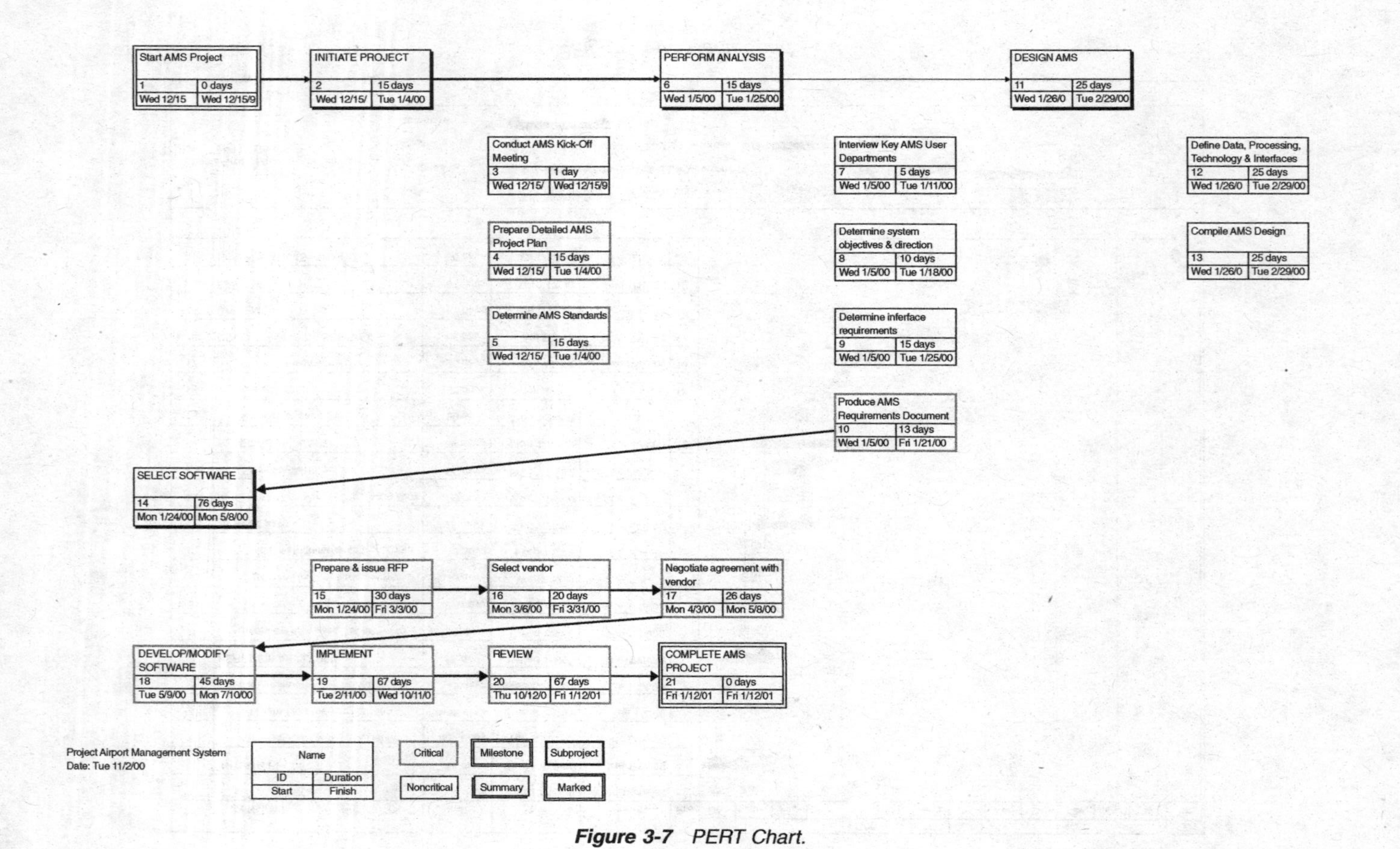

Figure 3-7 *PERT Chart.*

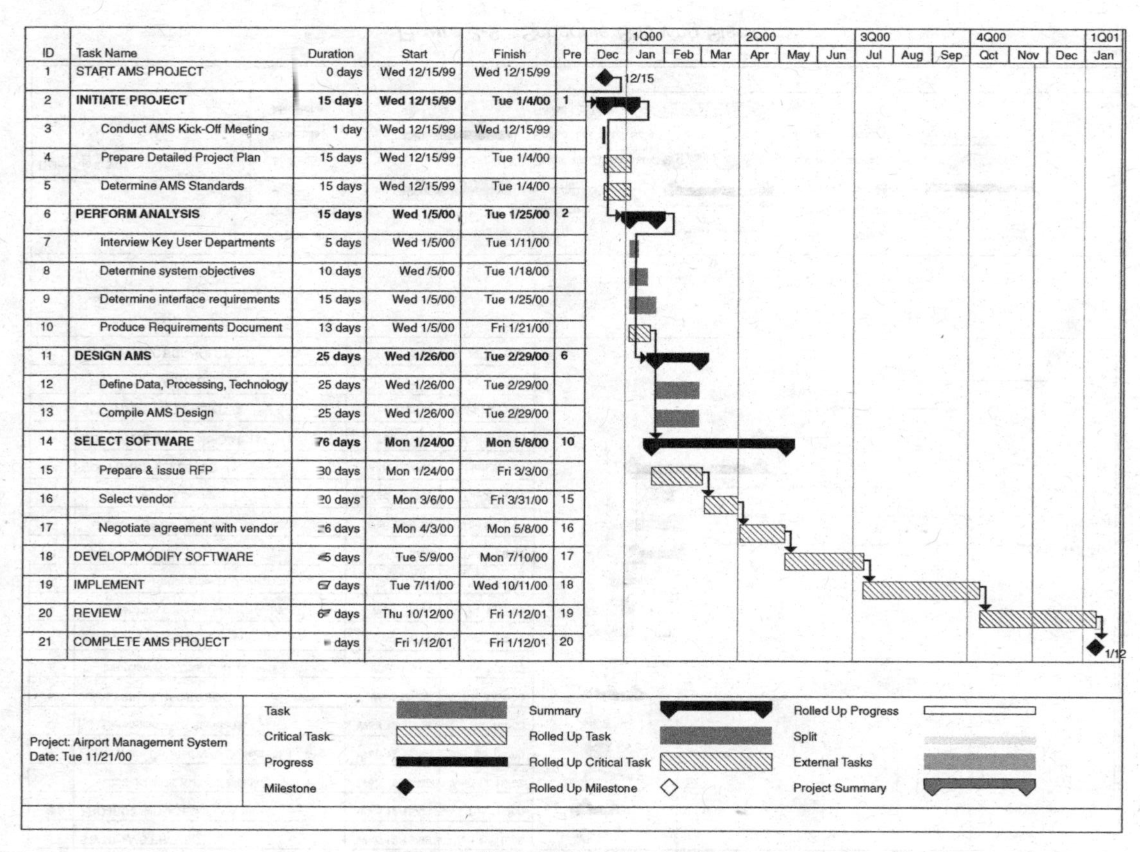

ID	Task Name	Duration	Start	Finish	Pre
1	START AMS PROJECT	0 days	Wed 12/15/99	Wed 12/15/99	
2	**INITIATE PROJECT**	**15 days**	**Wed 12/15/99**	**Tue 1/4/00**	**1**
3	Conduct AMS Kick-Off Meeting	1 day	Wed 12/15/99	Wed 12/15/99	
4	Prepare Detailed Project Plan	15 days	Wed 12/15/99	Tue 1/4/00	
5	Determine AMS Standards	15 days	Wed 12/15/99	Tue 1/4/00	
6	**PERFORM ANALYSIS**	**15 days**	**Wed 1/5/00**	**Tue 1/25/00**	**2**
7	Interview Key User Departments	5 days	Wed 1/5/00	Tue 1/11/00	
8	Determine system objectives	10 days	Wed /5/00	Tue 1/18/00	
9	Determine interface requirements	15 days	Wed 1/5/00	Tue 1/25/00	
10	Produce Requirements Document	13 days	Wed 1/5/00	Fri 1/21/00	
11	**DESIGN AMS**	**25 days**	**Wed 1/26/00**	**Tue 2/29/00**	**6**
12	Define Data, Processing, Technology	25 days	Wed 1/26/00	Tue 2/29/00	
13	Compile AMS Design	25 days	Wed 1/26/00	Tue 2/29/00	
14	**SELECT SOFTWARE**	**76 days**	**Mon 1/24/00**	**Mon 5/8/00**	**10**
15	Prepare & issue RFP	30 days	Mon 1/24/00	Fri 3/3/00	
16	Select vendor	20 days	Mon 3/6/00	Fri 3/31/00	15
17	Negotiate agreement with vendor	26 days	Mon 4/3/00	Mon 5/8/00	16
18	DEVELOP/MODIFY SOFTWARE	45 days	Tue 5/9/00	Mon 7/10/00	17
19	IMPLEMENT	67 days	Tue 7/11/00	Wed 10/11/00	18
20	REVIEW	67 days	Thu 10/12/00	Fri 1/12/01	19
21	COMPLETE AMS PROJECT	[illegible] days	Fri 1/12/01	Fri 1/12/01	20

Figure 3-8 *Critical path identified.*

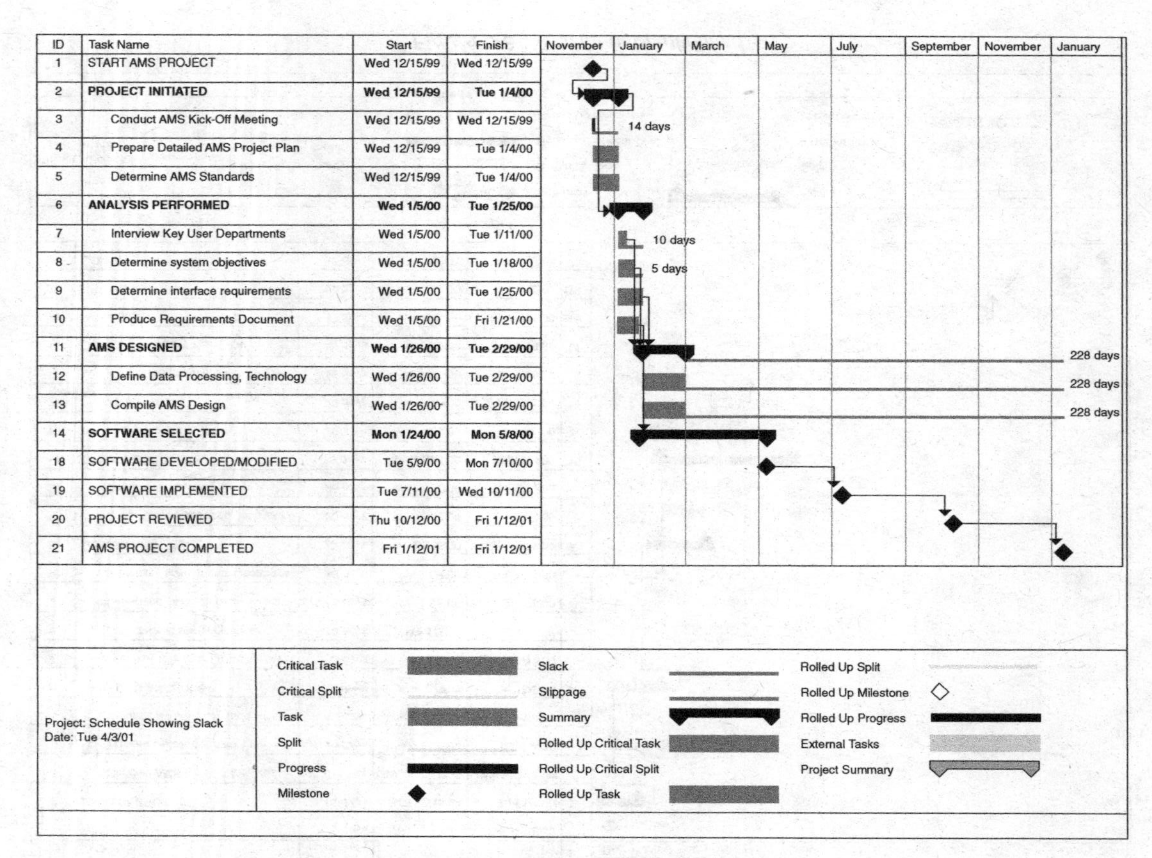

Figure 3-9 *Schedule Showing Slack.*

don't take up identical amounts of time. Any excess amount of time on the shorter leg is called *slack*. I could actually start and complete that task later, without impacting the schedule, because I can't finish the task anyhow. (Note: MS Project represents slack as a thin line.) Before automated scheduling systems, all of this had to be done manually. Now you can appreciate why people are so enthralled with automated scheduling packages!

One more term needs defining here: a *dummy task*. A dummy task is not one on which a person of little knowledge can work; rather, it's a "kludge" in the diagramming that indicates two activities for which no additional work is required but for which the arrows need to be connected. A dummy activity can be considered one that takes no time at all. You might want to use it to satisfy a precedent condition. It usually is represented with a dashed line to distinguish it from "true" activities.

Scheduling Summary

The best scheduling representation to use, in terms of your time and your team's understanding, probably will be the bar chart, in conjunction with precedence-based dependencies indicated. You will note activity durations on the bars, and indicate the milestone events on the nodes. Arrows tie the tasks together, so that you can tell, by dropping a vertical line for the current date, what should be happening at any point in time (and perhaps what has not happened when it should have). Refer to your project management software manual for more of the nuances, but "keep it simple" is a rule you should not violate.

I find that on small projects we seldom go back and redo the project schedule. While it's quite easy to do, the truth is that the other methods of project monitoring tell us quickly how well we're doing. The schedule reprinting only indicates what has happened; it does not identify which remedial actions need to be taken to bring things back in alignment—if that's possible at all.

Beware of . . .

Probably the most egregious errors I've seen in software project management were related to scheduling. Since a project is, by definition, a one-time occurrence, scheduling how long it will take is always a guess, at best—an educated guess, to be sure, using previous experiences of a similar nature, but nothing you can count on 100 percent. Every element of the performance specification can vary as you estimate the schedule on a project: late hardware and software deliveries, personnel out sick, unavailable servers, and so on. Here are some lessons I've learned over the years that might be of help:

- *Build on the experience of others.* When scheduling, include your team, your managers, and your peers, who may have had similar experiences of one or more of your work tasks.
- *Be conservative in scheduling external dependencies.* For example—and not meaning to bad-mouth the phone company—whenever I need a network circuit, I double the estimated time . . . and sometimes even that's not enough!
- *Be conservative in scheduling new technology.* Remember, leading-edge is also often bleeding-edge, so don't get "cut" when you're installing new software, new server models, or new cable. Even though it might have been successfully done elsewhere, each installation has its idiosyncrasies, all of which take time, and affect your schedule.
- *Be conservative in scheduling work from new technicians.* Be they programmers or analysts, if they're new to the application, the language, or the company, then add more time for them to deliver their product. One programmer with whom I had the pleasure of working for many years always took longer on the development side. Therefore, I always tripled his estimated time for completion.
- *Don't take on impossible projects.* If you are given a deadline, and after scheduling and analyzing the tasks, you determine there's no way you can deliver the project as required with the resources you have at hand, don't take on the project. (You can, however, first try to scale down the project or request additional resources, but even then your chances of success are not assured.)
- *Determine your project calendar.* Today's project management software allows you to create your own calendar, against which the schedule can be created. However, not all team members may follow the same time frame. For example, unions may limit work to a maximum number of hours daily, weekly, or in a given period; or refuse to work weekends. Balance these differences against the need for having separate calendars for your programmers,

who work 24/7 via the Internet, with the users who may belong to a clerical union that prohibits such scheduling. You'll then need multiple calendars for resources.

- *Don't forget to factor in vacations.* Especially on long projects—more than six months' duration—remember to allow time for your team members to take their vacations. Also, add in sick days, "Murphy's Law events," and holidays. If you don't have to use them, so much the better, but schedule them anyway.
- *Allocate time for additions to the project team.* If the project team members change, you generally will have to add time to bring the newcomers up to speed. An overly ambitious schedule precludes such an opportunity, in which case the project will get into schedule trouble because of the inexperience and lack of team knowledge on the part of the new members.
- *Practice rubber-band management.* When I schedule a project, I generally treat myself as an elastic commodity. That is, I don't add in the time it takes me to manage the project. I just work longer, if necessary, as there's no cost impact to my clients because I manage on a fixed-price basis. By now, I've learned that there will be random events that will keep me involved throughout the project.
- *Change the schedule when you change cost and/or performance.* Remember that the entire project management process is iterative. If you change what the software is going to do, or add to or subtract from project personnel or budget, revisit the schedule. You may be able to accommodate the changes within the schedule already developed, but more than likely you will not be able to do so, in which case you need to alter the schedule. (See Chapter 5 on monitoring for more information.)

COST PLANNING

The third project dimension is budget. Planning out how much the project is going to cost, and how much funding you'll need at each point in the project, helps you determine your real costs. Today's project management software allows you to determine your budget, and subsequently to monitor your costs, more easily than ever before.

Cost Types

Your project will have many types of costs: labor, for different categories of human resource; software and hardware; network services, plus rates added to those charged by your company and its internal accounting structure. Many project planning methodologies allow you to identify these different resources and budget for them while you are planning the project.

Laboring over Labor Rates

In my experience, the cost of labor usually is the largest component in a software project. Whether that labor is needed to install the network, configure the servers, customize the application software, input the database, train the users, or test the system, the fact is, you'll need people throughout the project. And almost without exception, you'll require people with different skill sets, and these differences are reflected in different rates for their labor. For internal projects, you can get your company's current labor rates from the human resources department, if they're not available elsewhere. On external projects, you can use recent project plans with similar labor categories to derive your hourly rates. However, as I noted earlier when addressing schedule, be careful in extrapolating someone else's estimates for your own purposes.

Typical labor categories on a software project might include:

- Systems analyst
- Programmer
- Data entry personnel
- Test analyst
- Network analyst
- Documentation analyst
- Trainer(s)
- Help desk personnel
- Project manager (of course!)
- User department personnel (managers, clerical and analytical staff)

You also need to determine who will defray the labor costs for user departments. This can be a sticky issue, as new software requires significant user involvement—and hours spent—for success. This effort can reduce the apparent productivity of the host department during development and implementation, despite the benefit of increased productivity down the road. The user department may have to defend its decreased productivity, and will want to charge back the hours spent to your project.

For example, in an enterprise MIS automation project, committees of 15 people spanning 30 departments met four days weekly every other week for two months to configure their parameter-driven system. At even a $50 hourly *burdened labor rate* (defined below), $96,000 was suddenly added to the project cost ($50 × 4 weeks × 32 hours weekly × 15 individuals). This can impose a considerable burden on the project, and put a crimp into the use of planned contingency.

In addition, you generally will have "burden" added to your labor rates. Essentially, a burdened labor rate contains a percent added to the actual hourly rate to accommodate the contribution made to overhead and/or administrative costs. So, for example, a burdened labor rate might be 140 percent of the nonburdened labor rate. This can come back and bite you during your estimating process if you've estimated only the nonburdened rate. For example, when you get charged for the labor hours in your internal accounting system, you'll see $56,000—or $16,000 more than the $40,000 (400 hours @$100, for a total of $40,000) you estimated! If you haven't estimated properly, you'll overrun your budget very quickly, without accomplishing the tasks necessary.

Another problem you'll confront is that often accounting systems, which calculate burdened payroll rates, don't hand off the data in a timely fashion (perhaps six weeks later), so you won't have a truly accurate cost status. In that case, you can do one of two things:

- Rely on your estimate of labor hours.
- Obtain promptly the hours-worked figures, reenter them separately, and calculate your own burdened dollar amount.

Still, in neither case will you have the precise cost, just a working estimate.

Hard Costing

In addition to your people costs, you'll have costs for hardware and software. These typically include:

- Telephone
- Travel
- Direct administrative costs, such as for supplies, media, and overnight deliveries
- Hardware (workstations, servers, routers, and modems)
- Application software
- Systems software, including additional licenses for server software and workstations
- Application software, indicating concurrent licensing limitations

Other costs might include chargebacks in the amount incurred, or with burden, for the following:

- Training rooms
- Test environments
- Documentation and training materials
- Quality assurance
- Help desk

The goal, using the WBS tasks, is to identify any and all costs you might incur. You then take those tasks and spread them out across the time you've scheduled for your project. This is important because your project will have a cost plan that will reflect the way in which the budget will be expended over the duration of your project.

If you've contracted with an outside source to deliver services, be sure to include in your contract the schedule of payments clearly specifying what amount is due your contractor upon the completion of each payment milestone (a task that, when completed, has money paid against it). (See the payment milestones shown in Appendix II.)

Synchronizing Cost Reporting

Each organization reports costs differently. If your accounting or finance department is reporting only biweekly, for example, then updating your records weekly with cost reports will waste your time. The cost figures won't change in the intervening weeks. Keep your budget reporting cycles and the budget breakdown to the level of granularity that can be readily supported. However, if you can obtain labor hours actually worked from your payroll system, you can calculate an approximate cost yourself. But, note, this is appropriate only where the risk outweighs all the effort you have to put in yourself.

PRACTICING THE ART OF ESTIMATING

If you take each task in your WBS, figure out which resources, both labor and nonlabor, will be required to complete it, and calculate the cost of each element for that task, it will bring you to a closer estimate of the cost for that task. The emphasis here is on *estimate*. Needless to say, there is no guarantee that the costs will turn out as you calculate, due to circumstances beyond your control—increases in mailing costs, charges for phone lines, to name just two. Your goal is to come as close as possible in your estimate, by including as many components as possible, so that the project plan is meaningful. Of course, the total project cost is the sum of the component task costs. A sample project budget, broken into component costs, can be found in Table 3-1.

In the public sector, most project budgets need to be approved by the governing boards or bodies. If the project is not properly scoped and budgeted, the body cannot award the proper funding for it. And if it is awarded, but then you, as project manager, keep going back for additional funding, questions regarding the viability of your project, or even worse, your management ability, may be raised. One way to prevent this, of course, is to always make such high estimates that you seldom have to go back for additional funding. However, that can also result in wasted resources, as project budgets tend to be consumed over time.

Improving Accuracy

Working with the WBS tasks, costing each one, and then summing them up is a bottom-up estimating process. An alternative is to do a gross estimate of the project budget, from a top-down basis. The top-down estimate is the amount derived when the project was conceived, typically, before the detail was planned out. In my experience, top-down budgets generally have little bearing on either the scope of the

TABLE 3-1 Sample Project Budget

Cost Category	Project Costs
Application Software	$134,000
Database Software	$4,000
Other Software	$33,000
TOTAL SOFTWARE	**$171,000**
Design Services	$34,095
Delivery, Installation, Configuration	$113,650
Software Development/Modification	$22,000
Data Migration	$40,000
Training	$34,095
Other Services	$5,460
TOTAL SERVICES COST	**$249,300**
Travel/other expense	$30,000
TOTAL SOFTWARE COSTS	***$450,300***
Year 1 cost	$27,600
Year 2 cost	$29,808
Year 3 cost	$32,193
TOTAL MAINTENANCE—SOFTWARE	**$89,601**
TOTAL 3-YEAR COST—SOFTWARE	**$539,901**
Workstations—15	$60,000
Servers	$12,000
Other Commodity Hardware—6 printers, 2 faxes	$8,000
Proprietary Hardware—routers, switches	$18,000
TOTAL HARDWARE COST	**$98,000**
Year 1 cost—hardware	$6,000
Year 2 cost—hardware	$7,000
Year 3 cost—hardware	$12,000
Monthly communications costs—3 years, $800 @ month	$28,800
TOTAL MAINTENANCE—HARDWARE	**$25,000**
TOTAL 3-YEAR COST—OTHER	**$123,000**
GRAND TOTAL COST	**$662,901**

ultimate project or its realistic cost. Not only do resource costs change over time, but, invariably, as the detail of a software project is developed, the cost estimates become higher. In rare instances, the costs decrease, such as when a commercially available software product can replace one that initially was thought had to be custom-developed.

For example, we had one client for whom $900,000 had been estimated for systems three years before the mega-project was even scoped. The estimate included everything related to technology: fax machines, network costs, phone line charges, and clerical assistance. The original number didn't reflect internal cost structures, chargebacks, or a common definition of what was to be included in the budget. But the $900,000 figure stuck in everyone's mind as the total funding available—which, after all the ancillary costs were subtracted, left inadequate amounts for the actual systems. As a result, the actual cost of the project, totaling $1.2 million, was perceived as overrunning the budget, rather than a reasonable cost.

As you gain greater experience in estimating, you'll be able to evaluate similar tasks to see if their costs are comparable. Perhaps there has been an incorrect assumption on the actual work to be done. This can occur when different people have estimated the tasks.

Parametric Cost Estimating

There is an old saying that people who don't learn from history are doomed to repeat it. *Parametric cost estimating* is based upon using historical data for similar projects as the basis for estimating the cost for your current project. Unfortunately, learning from history to estimate costs can be misleading. By using a statistical relationship between historical data and the variables, an estimate is generated. Similarities that might render cost information useful for similar projects could include:

- Project size. How big is it?
- Software application/business environment. What is its character?
- Level of new design and new code. How much new work is needed? How many new lines of code will be generated?
- Resources. Who will do the work? How experienced are they?
- Utilization. What are the hardware constraints?
- Customer specifications and reliability requirements. Where and how are these used?
- Development environment. What complicating factors exist?
- Complexity of effort. How difficult are the various modules?

However, even having the answers to all these questions is no guarantee that the estimate will be accurate; it can, in fact, be misleading if your input data have themselves been poorly estimated.

Experienced-Based Estimating

Experience-based estimates really do work best, for the simple reason that you can take into consideration what you have learned about the environment in which the project will be conducted and about the goals it is set out to achieve. As with schedule, costs will likewise be affected by these factors, both as to the achievable degree of accuracy and to the accuracy itself. The lesson here is to always use common sense when estimating, which means considering these factors:

- *Newness of the effort.* If this is a first-time development effort, obviously you can't know what it will take to get you to completion.
- *People working on your project.* New people will cost you more, in the amount of time they spend completing their work (as compared with experienced people who are typically more productive); you'll also need experienced people to mentor them. When you have human resources you've worked with, you can better estimate how long they'll take to accomplish a task.
- *Transferability of facilities and environment.* If you are using new hardware, architecture, systems software, or languages, the risk in your project increases. Generally, this will introduce greater cost to your project as well. If you're using the extant development platforms and language, you can better estimate how long it might take on this project.
- *Length of the effort.* In my experience, longer projects tend to cost more than originally estimated. This is often due to the difficulty of keeping productivity levels high for long time periods, or to the personnel turnover that naturally occurs in a project over time. You probably know whether your organization can respond to short deadlines, or if there is entropy built into the system that will cause productivity decreases partway into the project.
- *Project complexity.* Your company has probably shown you over time the critical level at which it can efficiently handle a complicated project. For example, it will take your staff longer to configure and install a system with three servers, as opposed to installing one server for three separate systems. Above that, there is a degradation in the performance of all resources. Once you know that level, you can better estimate your project costs in that area.

Using Outsourcing to Control Costs

One of the best ways to reduce the risk of overrunning the budget is to outsource the project, or pieces of it, for a fixed price. This increases your chances of bringing in the project within the stated budget. However, as in managing any internal resource, you must include the costs of managing the outside contractor in the budget. These include:

- Project or contract manager (internal)
- Quality assurance staff
- User training
- User design team
- Accounting services (for processing payments)
- Legal services (for contract review)

Caveat Estimator

When a cost estimate is done prematurely, often the project team has to live with poor results. You can minimize that possibility by working backwards; that means, take your WBS and, before scheduling the tasks, check to make sure you have adequate budget for all of the critical tasks. Perhaps you can do a partially phased implementation. For example, recently a client had preliminarily allocated $1 million for an infrastructure data management system. The costs of gathering the data were unknown, but after an initial assessment phase, the estimate to complete the project in its entirety nearly doubled the original estimate. Since this was leading-edge work, and the benefits of gathering and processing all of the data were unclear, we broke the project into an initial implementation for $.5 million. Depending upon the outcome of this initial implementation, the scope of the final implementation would be determined. At the end of the initial implementation, the client would still have a usable database with which to work, even if funds for the final implementation were not added to the project. We squeezed into the budget the tasks that could fit.

Labor hours, as I mentioned earlier, are typically used in planning. However, not all labor hours are equal—as anyone who has had to use a junior programmer when a senior analyst was called for can attest. Estimates of 4,000 labor hours in toto does not mean that you'll need only two analysts (at 40 hours weekly for 50 weeks each). Clearly identify the specific talents of the staff you need.

Finally, remember to alter the costs over time on long projects as appropriate. While today's inflation rates are nothing like the 15 to 20 percent inflation experienced in the '80s, you still can expect that your salary structure will show increases over time, and that your external contractors will raise their prices biennially if not more frequently. Consult with your peers and your company's finance department to get their cost growth estimates, then multiply your costs by that figure.

For example, if your project budget is initially estimated at $1.2 million over three years, and your hedge for the future inflation is estimated at 2 percent per annum, your actual budget should be submitted at $1,224,160, or $24,160 more, based upon a linear expense pattern of $400,000 annually:

Year 1	$400,000
Year 2	$400,000 × 2% + $400,000, or $408,000
Year 3	$400,000 + $400,000 × 4% or $416,160

CHOOSING A PROJECT COST SYSTEM

You will need a project cost accounting system against which to monitor your project budget. Your company's finance systems may have a project cost component, in which case you can use those numbers, as long as you understand their derivation and limitations, if any. Be careful of financial systems that carry annual operating costs only, not project costs, which can span multiple fiscal years.

Your accounting systems need to capture the information in such a manner that you can monitor effectively. That said, be aware that in most organizations I've seen, project costs typically have a "phantom" entry that comes from the finance department as a result of allocation of overhead costs among the rest of the organization. That's the overhead burden, or general and administrative (G&A) burden. Whatever it's called, it's clearly a burden on you—because your project budget will be diminished by it.

More important, you need to understand its role within your organization: not only will you have a burden on labor that you directly use for your project, but you will get charged back a proportion of the costs the company incurs to keep the business going—accounting, human resources, and so on.

There are four basic ways of arriving at the costs, all of which can produce different charge back amounts to your project. The basic components of your project costs are *labor* (direct and indirect), *overhead burden, nonlabor direct costs,* and *G&A burden.* G&A can be allocated based upon your department's revenues, or on costs plus revenues (less common). Overhead typically is the cost of running the department, less any revenues. Some companies charge a department, or project, *both* G&A and overhead. And because you don't have any control over the allocation, you must check with the finance department so that you can allocate the amount when you calculate your initial budget. Setting up a spreadsheet to note the costs or acquiring a job costing system will help.

I emphasize so-called phantom labor costs because they can ruin your budget estimates. As alluded to earlier, how you get charged back is a function of your finance department's process. You need to be aware so you can plan properly. For example, ask:

- Are special charges made for material handling?
- What costs comprise the G&A? Do staff bonuses go into it?
- What costs are computed as part of overhead?
- How frequently are they computed?
- What is allocated in proportion to your expenditures, and what as a single sum?
- Especially in technology-related departments, everyone uses your services. Do network maintenance and new network infrastructure costs go into your G&A? Your overhead?

A critical element of project cost accounting is to be able to prohibit unauthorized departments from charging to your project. Newer project accounting systems limit such entries to authorized people. However, errors can still be made, and you will have to spend the time to figure out where the money went—and, if necessary, how to complete your project within the original budget!

Holding Back

Did your family ever have a dinner party to which more people came than had been invited? Your mother probably told you to hold back from eating too much until all the guests had taken their share, and you could be sure there was enough for everyone. Well, when you're managing projects, and especially when you're competing with others in your company for the same set of resources, you may be put in the situation of having to take a back seat and wait for critical resources your project requires while other projects take "their share" first.

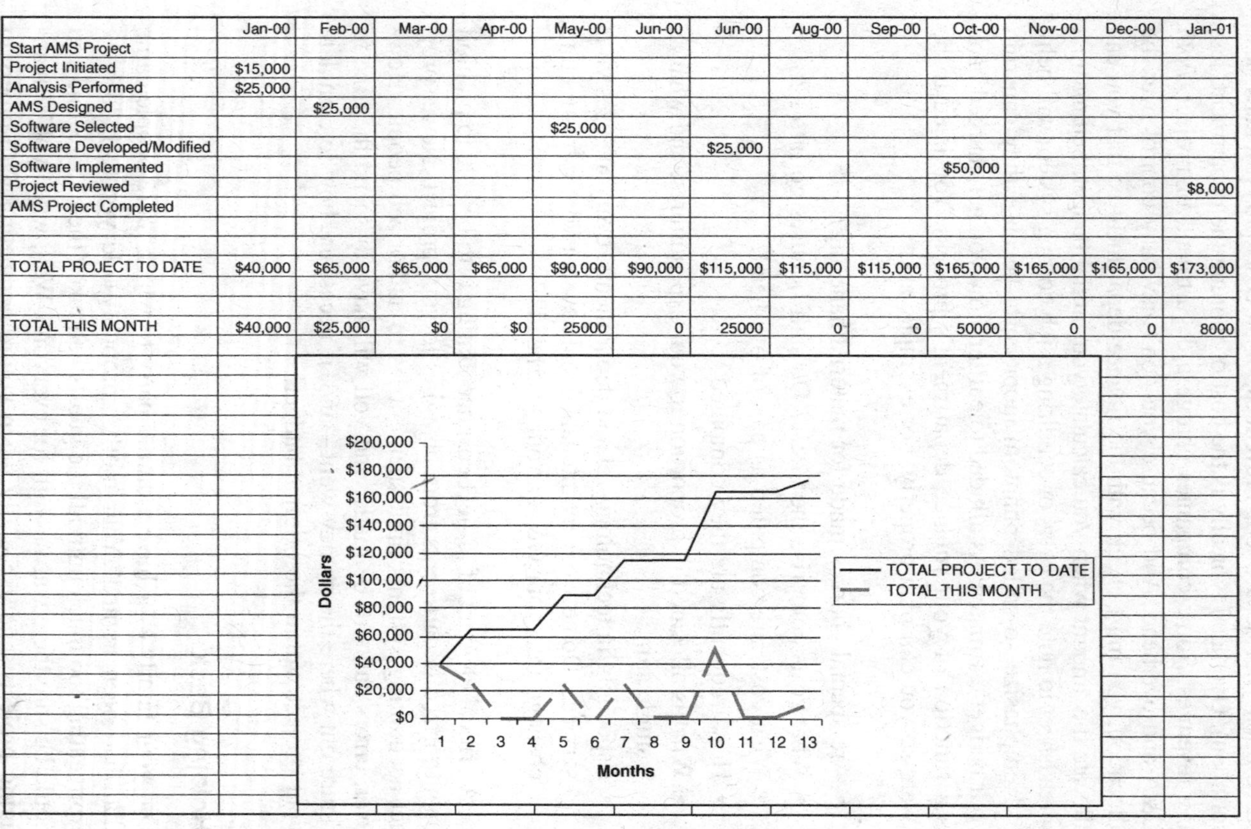

	Jan-00	Feb-00	Mar-00	Apr-00	May-00	Jun-00	Jun-00	Aug-00	Sep-00	Oct-00	Nov-00	Dec-00	Jan-01
Start AMS Project													
Project Initiated	$15,000												
Analysis Performed	$25,000												
AMS Designed		$25,000											
Software Selected					$25,000								
Software Developed/Modified							$25,000						
Software Implemented										$50,000			
Project Reviewed													$8,000
AMS Project Completed													
TOTAL PROJECT TO DATE	$40,000	$65,000	$65,000	$65,000	$90,000	$90,000	$115,000	$115,000	$115,000	$165,000	$165,000	$165,000	$173,000
TOTAL THIS MONTH	$40,000	$25,000	$0	$0	25000	0	25000	0	0	50000	0	0	8000

Figure 3-10 *Cash Flow.*

I've seen development projects for which critical resources were not available when needed, and the project was put on hold until the resources could be freed up. Unfortunately, when those critical resources were finally freed up, some people, necessary to the project's successful outcome, had left the organization or had become involved in other efforts. You can use contemporary tools, such as MS Project, to perform *leveling* of critical resources for you. But note that all of the projects using that resource must be available for the tool to properly do a resource-leveling calculation. Clearly, you need to be able to "roll up" multiple projects in order to forecast workloads and demands on critical resources.

Spreading the Wealth

Generally, a project spans many periods—months, quarters, and years. When financing for your project is issued on a periodic basis, you may find that your planned expenditures don't gibe with the availability of funds. This is where a *time-based cash flow* is handy. You can see how much your project will need at critical points in the project. For example, if funding is available at the start of each fiscal year, you want to be sure you don't go over budget specifically at that time. Running a quick cash flow report will show your cash needs by month or quarter. If you find yourself getting perilously close to running over budget, you can slip the schedule, and the cash resource needed, to make sure you don't end up in a sudden budget crunch.

A spreadsheet that shows planned expenses over time can be quickly turned into a graph, showing you where the belt needs tightening. A sample of this is shown in Figure 3-10. Note that expenses are highest in month 10. If that conflicts with other expenditures, you might want to adjust the schedule or break down the task into multiple payments to eliminate budget pressure.

BUILDING IN CONTINGENCY

All project plans should have contingency built into them. Because the best-laid plans can, and usually do, go awry, you would be well-served to build contingency into both your schedule and your budget (since budget is generally affected by delays in your schedule). In my firm's software projects, especially for those in the public sector, we build an

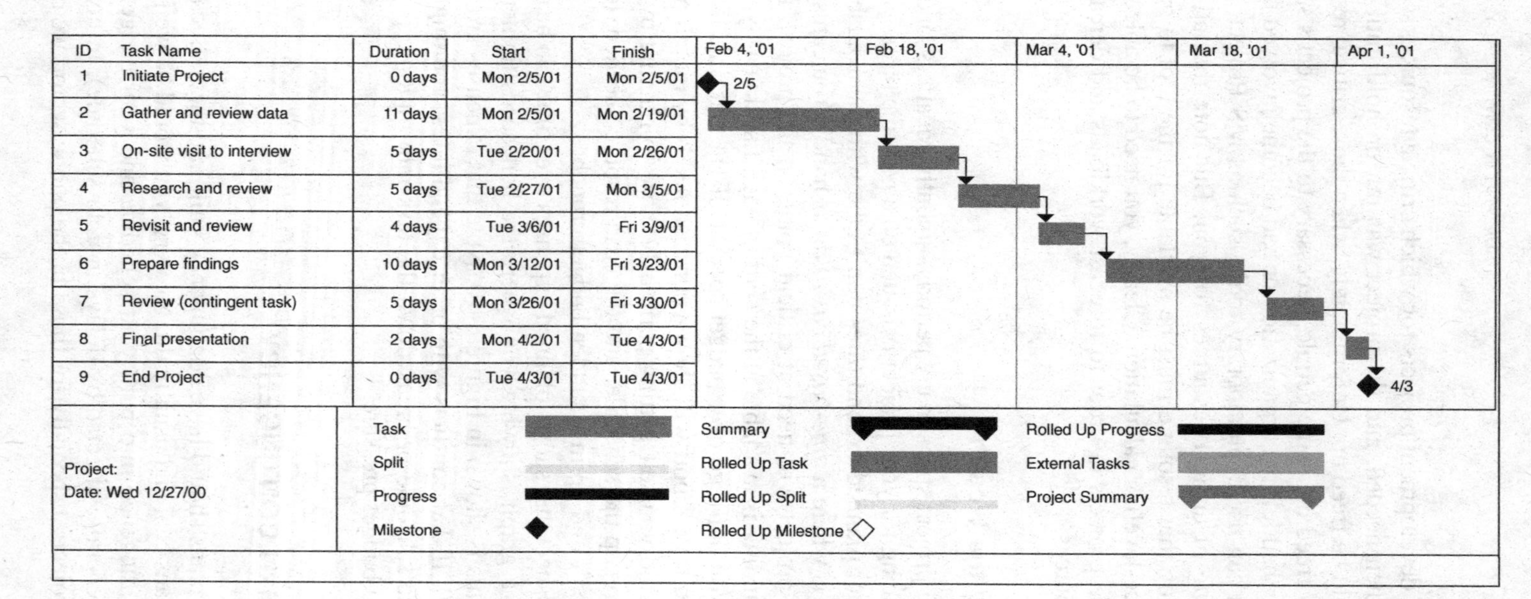

Figure 3-11 *Contingency in Schedule.*

overall contingency into our initial request for funding. That way, we can issue changes within that contingent amount without having to go through extensive rebidding and reapproval processes for funding within the foreseen amount. Generally, we plan for 10 to 15 percent contingency, depending upon the project complexity (as explained in Chapter 2).

But there are two problems with contingency that you should be aware of: it can be no more accurate than the underlying schedule and budget calculations, and it is not a substitute for the planning you and your team need to do. Having said that, however, you can "backload" your schedule with an additional task toward the end of the project, as noted in Figure 3-11. The purpose of task 7 is to accommodate any "cracks" that have developed in earlier tasks and need repair.

Of course, these software cracks—such as the need for additional testing, subject matter experts to resolve a recurring network problem, and an additional piece of software—cost money and take time. Pre-identifying the separate task reduces your chances of being under budget and behind schedule more than other contingency calculation methods, such as adding a predefined amount of time and money to each task. In that case, you've taken your estimates, padded each to make them all more inaccurate, and introduced inaccuracies into those tasks for which you had a measure of precision.

RISK PLANNING

In general, there are three steps to take to reduce risk to a project: First, you identify the most likely risks your project faces, based upon the political, technological, technical, managerial, and personnel unknowns; second, you assign each one a likelihood of occurrence; third, you decide on a method for mitigating the likelihood of occurrence.

For example, if you have an inexperienced staff, the likelihood of discovering bugs during unit testing probably is 100 percent. Mitigating methods might include additional unit testing, mentoring sessions by more experienced programmers, and the replacement of one or more juniors with an experienced programmer for the most critical software modules. Then you assign a number, such as:

TABLE 3-2 Risks and Likelihood of Occurrence

Potential Risk	Likelihood of Occurrence	Impact of Occurrence
Inexperienced programmers may cause lengthy and costly testing.	3	C
New operating system release may create integration problems.	2	C
Financial system vendor will not support product any longer.	3	D
Customization of accounting package may create integration difficulties with other products.	3	C
New system will not be supported by users, due to personnel turnover, with no one here when system is finally implemented.	2	C
Highly volatile rate calculations may cause inaccurate invoices.	1	D
Hardware vendor will go out of business.	2	B
Purchasing system will not be ready for user training as scheduled.	2	B

Probable = 3
Possible = 2
Improbable = 1
Impossible = 0 (Why is this on the list in the first place?)

I plan for the 3s and 2s, which are the most likely. And for each of those, I evaluate the impact of their occurring and assign another ranking. For clarity, I like to use letters:

Catastrophic = D
Difficult = C
Inconvenient = B
Irrelevant = A (Why is this on the list in the first place?)

The situations to plan for are the 3s and 2s that are also Ds and Cs. Anything ranked lower can be dealt with if and when it occurs—unless, of course, you have such resources, including your own time, that allow you to develop plans for the impossible. Recall Y2K software issues: even for those, not every possible situation was accounted for in corporate contingency planning, just those that would be most critical if they did occur.

You can't always predict which unforeseen events will occur, but you can plan for them, so that when they do, you can swing into remedial mode promptly. My method is to prepare my plan, review it, and identify those events most likely to occur. Then I look at the threats to my plan, and come up with solutions or action plans to mitigate those threats or to deal with the unwanted situation should they arise. A sample list can be found in Table 3-2. You will come up with your own list over time, but this is a good place to start.

4

Lewin on Leadership

Perhaps your most important task as manager of a software project is leading your team members. Well-developed leadership skills will enable you to motivate your team to work long past quitting time when necessary, and will inspire them to generate and sustain the adrenaline needed to work nonstop during the testing and implementation phases. After more than three decades of project management, I have seen projects succeed that everyone thought would fail. Conversely, I've also seen projects whose success appeared guaranteed, fail—the Quadruple Constraint was reasonable, the risk assessment was low; the team members were talented and appropriate for the task. The weak and fatal flaw in the projects that failed was inadequate leadership. This problem is more acute in project organizations than in line organizations because a project manager can influence an individual's future in a less direct fashion. For example, in many organizations, the project manager can, at the completion of the project, write an evaluation of performance. However, the line manager may not include it in his or her evaluation of the individual. And in some organizations, not even that can be done.

I perceive the leader of a project as the rudder on a ship, guiding his or her team through the narrow waters of political intrigue and personnel changes, through the mishaps of hardware and software snafus, and through the straits of user misunderstandings and demands. Leading a project team is, in my experience, much more difficult than leading a functional department or group. In the latter case, evaluations are done on a regular basis, the reporting structure is more static, and the goals are clearly aligned with the enterprise's goals. A project team, on the other hand, has typically been assembled for this one endeavor, after which each member returns to his or her own department, to await another project or departmental assignment. The project leader, unfortunately, has little authority over where the next assignment will be.

ORGANIZING THE PROJECT TEAM

The project team is your human resource that will assist you in taking the project from inception through completion. It will comprise any number of people from different parts within the organization, and its active composition may—very likely, will—change during the life of the software project. (Note: Anyone who participates in your team is considered part of the team, even if another department administers that person.)

We include users in our project teams extensively, as we've discovered that only users can successfully implement a system in their environment. On our recent airport project, for example, our team comprised the following:

- Users from airport division management
- Users from critical internal groups (both supervisory and clerical levels), such as operations, development, and facilities
- Technical network expert from the IT department
- Finance department account manager
- IT program office
- Audit and risk managers
- Vendor developers and account manager (after selection of the software product)

Individual participation varied over the life of the project, following that shown in Figure 1-5 earlier. The program office served as facilitator and coordinator, but the project team leader was, ipso facto, the leader. The program office wrote documents, raised issues, and helped the project team navigate the maze of technical issues as the project took shape; but, the project team leader was the critical resource: she set the pace, held team members accountable for assigned activities, and insured that internal departmental needs were met. She also introduced a project management methodology to the project to make sure that all resources required, both internal and external, were accounted and planned for.

This type of departmental user organization, with expert assistance, is particularly good for situations where packaged software will be implemented, and where users already have a system. The users know what they want better than anyone else, because they're familiar with the problems at hand.

User-Led Team Issues

As a consultant, I prefer user-led teams. The users know best the political and communication paths to take within the organization, and are best situated to identify potential problems before they arise. However, problems may still arise because the leader might not have worked on a software project before, let alone led a team. Problems I've noted that can arise are as follows:

- *Collateral duties.* Users often are assigned to a project team without being relieved from their normal tasks. For example, in a recent enterprisewide task force, the technology purchasing agent was also on the project committee. His purchasing tasks suffered because he had to attend so many meetings, and no one had been assigned to back him up in the purchasing department.
- *Inexperience in leading meetings.* While good intentions abound, first-time project leaders often don't know how to make meetings productive. Rather than arrive at decisions, meetings become gripe sessions, or are, simply, unfocused; specifications give way to subjective judgments. In such situations, users often give up and drop off the team.
- *Subjective judgments.* When selecting packaged software, some users prefer a particular vendor. For example, they value salesmanship over substance. And, when the development team replaces the salesperson, who is never seen again, user enthusiasm is often dampened. To get through this, an evaluation matrix, such as that shown earlier in Table 2-2, is recommended. Team members should agree to the evaluation criteria first, then complete the matrix for each candidate.

- *Lack of quantification of training.* Training is imprecise because individuals learn at their own speed, despite the teacher. It's often hard to determine whether attendees at training sessions really understand the importance of what they're attempting to learn. Thus, an evaluation instrument, such as that shown in Figure 4-1, quan-

The purpose of this evaluation form is to determine how well the training satisfied your needs. Subsequent actions will be based upon your response, which is appreciated.

Please complete and return this form by Thursday close of business, 1/27/02, to Marsha Lewin at marshalewin@acm.org or via interoffice mail to Marsha Lewin.

---

1. How would you rate the training, assigning a number from 1-10, where 10 is best and 1 is worst, on the following aspects:

   _____ content _____ pace of instruction

   _____ facilities _____ relevance

   _____ overall instructor effectiveness

2. Do you understand your role as department trainer?

   Yes_____No_____

   If No please indicate what areas require further understanding:

3. Do you have questions, for example, how the Contract Management Product applies to your department's needs?

   Yes_____No_____

   If Yes please indicate what these questions are:

4. Do you want to form an internal trainers' Users Group to meet on a continuing basis to resolve interdepartmental and intradepartmental issues such as: to identify sharable codes and databases, common procedures, training materials, and standards for field labels; to streamline screens; and to answer questions that arise?

   Yes_____No_____

   If No please indicate why not:

5. When will you initiate the use of the system in your department?

____

Figure 4-1 *Training evaluation form.*

tifying understanding of the materials is very helpful. Not only does it evaluate the material and its presentation, but it can also be used to improve subsequent training sessions.

- *Lack of rigor in change control.* Often, users don't understand the need for controlling changes in an orderly fashion. Users tell the developers what else they'd like to see, or where to tweak what they've presented, and the obliging developers spend countless hours, even days, producing something that bears little resemblance to the software expected by the rest of the organization.
- *Inability to think "out of the box."* Users, especially those who have been in the same organization for a long period of time, often have difficulty visualizing how the new software will change the current processes and procedures. As a result, they may evaluate or design the software to replicate existing inefficiencies in their systems. The team leader should act as the visionary, to help the team imagine how new tools can change their roles and responsibilities for the better.
- *Inexperience planning along the Quadruple Constraint.* It goes without saying that planning budgets and schedules is easier when you know your resources and have worked with them before.
- *Failure to keep line managers informed as to time commitments and progress.* Since user teams are formed at the behest of their sponsoring management, the team leader must keep these managers in the information loop so that they are up to speed on activities that may require more time—or less time—over the life of the project. This prevents the project team leader from having to renegotiate the time commitment with each manager through each phase of the project.

KNOWING WHO TO ORGANIZE

Where do project team members come from? You already know that often they come from user departments; this is always the best way, but it's not always possible because today organizations run leaner and meaner, resulting in fewer available resources for project organizers to choose from. So now is the time to bring up outsourcing, the use of external resources to perform all or part of your project's software development.

Going In-House or Out-of-House?

Increasingly, I see organizations going through an RFP process and outsourcing the entire effort. The in-house users help only with the requirements determination and during the final testing. This way, the organization does not have to support a large internal development staff. The internal/external involvement is divided as shown in Figure 4-2. Though certain tasks still must be shared, a clear division is drawn between many user and outsourced tasks. Note that additional project management is required with this model, as the vendor will have its own project manager to monitor its resources, while you will safeguard your company's interests. Unless someone is managing the project internally, you may be faced with erroneous invoices, resulting in payment for services *not* rendered.

In spite of requiring additional management, outsourcing software development has a great deal of merit in this era of multiple platforms, languages, and delivery systems. By outsourcing a project, an organization does not have to hire expensive and difficult-to-keep technical expertise. It also doesn't have to spend money reeducating its own personnel. The emphasis is on training the users, who ultimately are the ones who have to know how to work with the system.

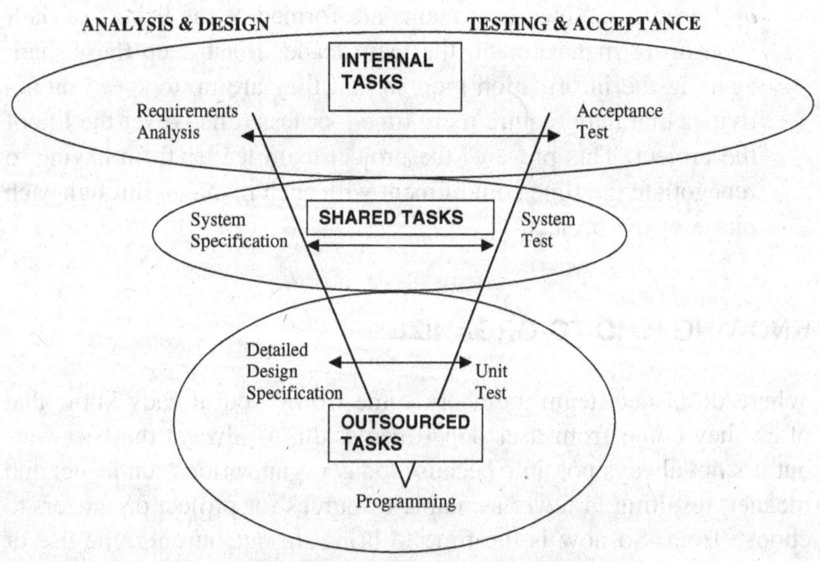

Figure 4-2 *Outsourcing of tasks.*

The secret to outsourcing without causing more problems for yourself is to ensure, as stated earlier, that the contract, or performance specification, is unambiguous and complete. If you outsource software maintenance, a *service-level agreement* (SLA) is also appropriate. The SLA must specify response times, performance criteria, and responsibilities, to ensure that not only will the project be feasible to accomplish, but that both the outsourced company and the customer organization will have incentives to perform well.

To describe the internal software project team, let's assume we're developing software. The team should comprise at least the talents represented in the organization chart in Figure 4-3. Certainly, one individual might be able to serve multiple functions during the life of the project, but generally you won't have that luxury. People with unique talents just are not available to do lower-level tasks; their talent—such as network design—is in demand on other projects. Even though they might be very helpful in documenting the system, they'll rarely be available to do more than contribute their expertise.

And that brings up the question of support staff, people who do ancillary jobs that are required for completion of your project, but who do not report directly to you—for example, computer operations staff, network staff, documentation and software services, or as they're called in some organizations, the help desk. You'll need all of these resources in order to complete your system. But these individuals generally are part of an ongoing departmental organization outside your project control. This means that if your project is not on their radar screen when you need them, you will experience significant delays.

For example, in the police system my team installed recently, we needed the network expertise from the IT department to connect mobile computers in patrol cars to and through the state's Criminal Justice Telecommunications Network. IT had enormous lists of other tasks, and could easily have put us on the back burner and delayed our project. But because we got on their list early, and stayed in communication with them as the critical dates approached, when we were ready, they were, too. They had been given prior notice, and had been reminded, and hence, were not taken by surprise. And, of course, after they'd made their contribution to the project, we thanked them and reinforced their importance to the project success.

The point here is that your team will be made up of many individuals with different talents. Some of the team will stay with you throughout, and others will be part of the team for short periods of time. Still

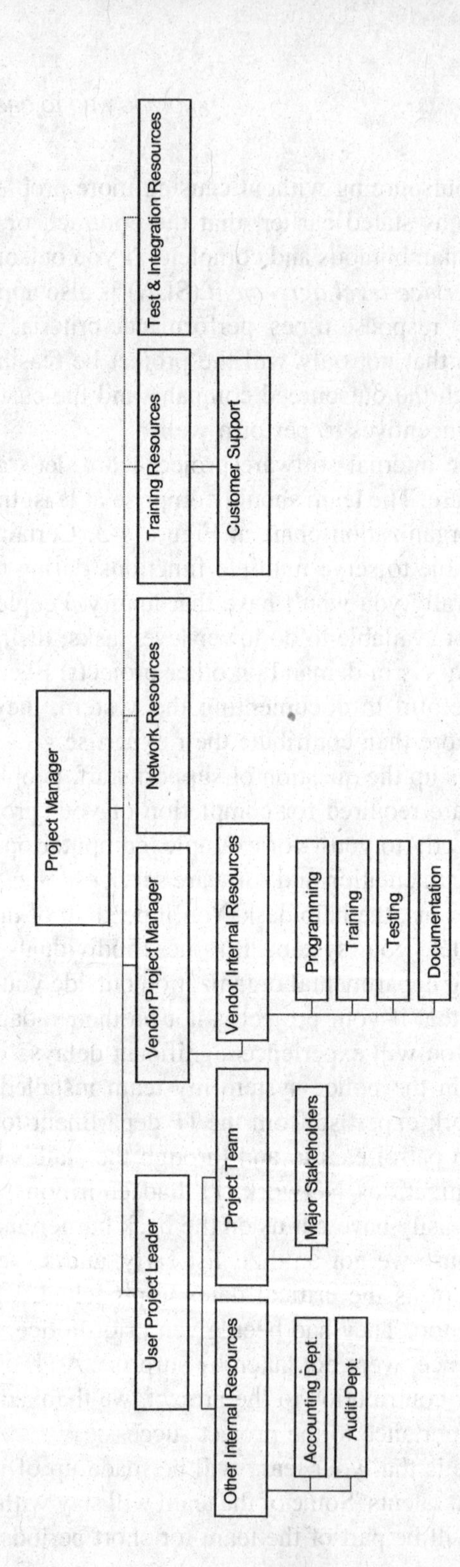

***Figure 4-3** Project Organization Chart.*

others will never be part of the team itself, but will serve to *support* your project. No matter what their role, you must communicate with them, so that they can do their best for your project.

One final word about the importance of communicating: Don't forget to be clear when you assign a task to your team members. Meeting notes are an effective way to make sure each team member gets the same information. Appoint a scribe, who is also responsible for distributing the notes after each meeting. Identify action items to each individual. Write down (or use e-mail) as specifically as possible what action is required, and by what date. Then, if you subsequently have to reassign a task, for whatever reason, all you have to do is cut and paste. Of course, finding someone else to assign it to will be more difficult!

LEADERSHIP GUIDELINES

While I haven't yet encountered the Pied Piper leading his or her team off piers to watery doom, I have seen managers who started with motivated teams, then inexorably ground them down with poor leadership. A number of lessons I've learned from these examples might help you to avoid making the same mistakes:

- It's *not about you; it's about them.* One manager I remember lost interest, then staff, and, finally, much-needed support because he was worried more about his own political value than his team's welfare. He ignored his team's personal problems, intraoffice issues, and scheduling conflicts. Ironically, when his staff resigned, and the project failed, he lost any political clout he might have had.
- *You set the pace.* The best projects to work on have a sense of urgency and commitment, and the project manager imparts those qualities. If you don't seem to care, then your team will not. If you doubt the success of the project, your team will assure its failure.
- *Reinforce the Quadruple Constraint with your actions.* If you insist that schedule is most important, but repeatedly delay in responding to the team, your example tells them that time isn't that important. Pretty soon, they'll be slowing down their deliveries, and the project will fall behind schedule.

- *Run the political offense for your team.* Especially on cross-functional and cross-departmental teams, you'll have conflicts to resolve. That's particularly true today, as software becomes more enterprisewide, to achieve process reengineering efficiencies. Departmental managers tend to defend process turf from shrinking data fiefdoms. Proactively address these issues with the affected managers, to be sure your project succeeds.
- *Feed the rumor mill with truth.* If you don't publicize your project, others will—with inaccuracies and innuendo. So be sure to maintain enterprisewide visibility on project status—major successes as well as interim achievements, even glitches.
- *Be positive.* The best project managers I know act as though there's always a silver lining in the clouds. They keep on moving forward; they don't lose faith in the outcome. No matter how many bugs you discover when you test, it's that many fewer bugs that would have been discovered *after* implementation. You can look at it as too many bugs, or as good testing. The team typically reacts better to positive reactions, even in the face of bad situations.
- *Support your team members.* I recall a project where some team members were caught in the middle of turf wars between their managers and the IS department, the sponsors of the project. Team performance suffered, as did the project. Approaching the managers and clarifying their concerns released the members from the political vise, and the project went on to achieve success.
- *Be on the lookout for burnout.* Especially in projects of longer than six months' duration, even the most energetic of individuals can fall prey to the stress of overload. In such cases, enforce vacations. Both the health of the individual and the health of the project will be the better for it. Be prepared for resistance, as one symptom of burnout is the inability to let go of responsibilities. Hold fast to the need for a hiatus.
- *Empower your team.* That doesn't mean let them make decisions without your following up. It does mean checking to see if they're proceeding according to the decisions they've made, or if they need help in their decision-making processes. With you backing them up, they'll better support their own decisions. And, in the longer term, they'll learn to make better decisions.

TALENT SEARCHING

Now that I've explained the importance of strong leadership, I'd better tell you where to find the talent you need when you need it. For software projects, there are three main avenues to pursue: seek "inside" sources; train your own staff; search the Web. Let's explore them one at a time.

Seek Inside Sources

An often-overlooked pool of talent is from the team who worked on the project proposal. In my experience, however, most organizations want their proposal writers to keep proposing, so rarely are they available for actual project work. And in situations where outside vendors have been awarded a contract, generally, their system installers have been named in the actual proposal, and will be part of the project implementation team; but others may have contributed only their resumes to the proposal.

My project teams typically are made up of the company's employees, strongly represented by user departments, with various technical talents added according to the particular project's needs. This way, we have people who are familiar with how the company does business, and we can use their past experience to our advantage—they already know to whom to address questions within the organization.

Train Your Own

When you can't find the precise talent you need within the organization, you may be able to find willing students. For example, in our contract management system, we had contract administrators who understood inspection, but not the technology. We sent them to classes in the internal systems of the company. Though we spent more on training than we might have if we had used others who already knew the technology, these administrators understood the company processes, and so were able to contribute good ideas for system efficiencies. The project turned out more successful, and was implemented more rapidly, than might otherwise have occurred.

If you decide to go this route—train your team—remember, you need to allow time and budget for both the training and the staff's learning curve. You will also need an additional mentor resource.

Search the Web

On occasion, when I've needed a specific talent for short periods, I've had success acquiring help via the Internet. I recommend this option, especially if you need specific talent, such as an Oracle programmer or a UNIX-to-NT conversion consultant, for a limited period of time. Just be sure to ask for references, and, specifically, confirm that there are no conflicts of interest. In the latter case, it would be embarrassing to find out that your newly hired Net talent had a day job working for one of your competitors.

In the software development profession, using remote resources for programming and for such tasks as documentation has been in vogue for decades now. In the eighties, I recall technical resources from India who programmed in COBOL. The results were sent via overnight mail as tapes or disks; and, in later years, the results were uploaded to customer sites that could accommodate file transfers (FTP sites).

Since names and e-addresses of such resources change frequently, I suggest you use a search engine to locate the talent you need, and choose from the results.

My firm has been using Internet technology for many years for common development, eliminating the need for on-site space for all of the development teams. We always use it for project communications and project management. I no longer hire anyone for our project teams who cannot communicate through e-mail, whether on- or off-site. Our documents, scheduling, and tracking are electronic, so anyone without such capabilities would slow down the decision-making and communication processes too much to accomplish the project's goals.

Electronic Security

An aside about remote access to your development team is in order here. In today's networked environments, firewalls are in widespread use to ensure that network security is not compromised by hackers. One of the major issues that should be addressed by any project planner is how the external members of your team will communicate during and after the development of the software. Common solutions are Citrix servers or PCAnywhere, which allow file access and manipulation by outsiders (hopefully, only to those authorized). Most vendors require such access for ongoing support of their software.

You can establish a restricted access site on your company's Internet to accommodate external contractors and companies; or you can use your corporate intranet so that members of your team can be assured confidential communication of project information. You can also publish project status enterprisewide on an unrestricted basis.

Be aware that if your company does not already support such external access, implementing it will become a mini-project of its own, and resources must be allocated for this infrastructure to ensure that firewalls are not compromised. Furthermore, depending on the network and e-mail organization, you may also need to create accounts on your network for your external—and even internal—project staff. There generally are rules for doing so, which are enforced by your IT department to ensure that only authorized individuals share project data. However, especially with a team from many different departments and organizations, the rules may cause conflicts, so be sure to check with your network support staff to minimize technical communication problems later in the project, when you least can afford them!

ORGANIZE A VIRTUAL TEAM

I touched on the topic of virtual teams in an earlier chapter, but I want to go into more detail here. I've used virtual teams for more than 15 years, since e-mail allowed easy access through telecommunications. To reiterate, a virtual team is one where the individuals comprising the team are not collocated, although they may physically convene on occasion. Communication is through electronic means, often using, in addition to, e-mail and discussion groups, collaborative tools, such as Lotus Notes, to create and revise documents.

Obviously, one benefit of using virtual teams is that it greatly reduces travel costs. For software development, which has clear specifications and tasks that can best be performed by individuals who require little interfacing with others, virtual teams work extremely well. Moreover, integrating external vendors and contractors into your virtual team can further expedite your project.

Tips for making virtual teams work effectively include:

- Minimize the interfaces with others when you assign the tasks. For example, make one person responsible for programming an entire module, rather than splitting the programming among many people.
- Keep the time for delivery short. That way, if a team member is in trouble, you'll know quickly and may have enough time to remedy the problem.
- Build your team using professionals you've worked with before. They are familiar with a virtual work environment, and you can more accurately schedule their time. You won't have to spend your time (and theirs) bringing them up to speed on the process.
- Use individuals who work well alone, and don't require the stimulation of others in close physical proximity.
- Make sure your sponsor is comfortable with an "invisible" staff. This is less a problem today with most people connected via e-mail in homes and offices. But more insecure managers often need the reassurance of seeing "their" people at the office, regardless of the cost to the project.
- Be very clear about project assignments, reporting structures, and reporting methods. Indicate clearly to whom each virtual team member should go with questions.
- Use e-mail liberally. And make sure that you, as the project manager, are copied on *all* e-mails. For you this will mean a full e-mailbox, but it will keep you apprised of how all your team members are progressing, and where they may run into snags and need future assistance. Also, use your e-mail's Forward capability; or use Novell's Groupwise or a similar product to set meetings, delegate assignments, record tasks, and track due dates.
- Where possible, have an initial in-person meeting with the team, to break the ice and to make future communications easier. To be sure teleconferencing is an alternative, but the technology gets in the way of getting to know one another on a more personal level. This face-to-face meeting will help establish deeper relationships of trust that will reap benefits during tight spots later in the project.

- Know when to step in—in person. Sometimes, the e-mail technology can cause problems, too, and you may need to spend some one-on-one time with one or more team members to work through an issue or to brainstorm a solution to a problem.

ADDITIONAL NOTES ON TALENT

If you cannot find the human resources you need when you need them through the three main venues I've discussed, you might have success the old-fashioned way: use temporary agencies that provide technically trained professionals. Often, companies have established agreements with one or more such temp agencies, which can provide resources more quickly than is possible through internal human resources (HR) departments.

Be prepared, however, to pay more than you otherwise would to obtain critical subject matter experts (SMEs), no matter the source. But if you tightly craft your agreement with such SMEs, you can use their time wisely and add value to your project—without unnecessary cost.

However and wherever you find your staff, make sure that you are indeed acquiring *talent*—individuals capable and desirous of working on your project. Especially on user teams, I have found that some individuals being released by their departments to serve on these temporary projects are those who provide the least amount of value and benefit to their own departments. Their managers, understandably, don't want to give others their better performers, so you may be forced to blend less qualified staff into your projects. Generally, such individuals rarely contribute much at the project team level either. Moreover, their inaction and, often accompanying, negativity can infect your project team, rendering the entire team less effective. In some cases, you may be better off doing without such "donations," especially on a project of short duration. On longer projects, where the user contribution is essential, it's best to negotiate with the department manager to get someone else.

BACK TO THE DRAWING BOARD

Staffing issues may cause you to realize there is an overwhelming need to replan: team members have departed; unplanned corporatewide reorganizations have affected who your system users will be; or new management has been unable to come to a consensus. These are just a few situations I've encountered. There's nothing wrong with replanning. Just make sure you're not constantly replanning to address small issues, in which case you're not really replanning, you're reacting to the normal random events that make projects lively and exciting!

When you do need to replan, however, be sure to communicate the changes, then process the changes through the change control process, as described in Chapter 5.

5

Monitoring

As the project progresses, the idea is to monitor, or control, what is happening, to avoid surprises in one or more aspects of the Quadruple Constraint. Control, by the way, is not a nasty word: that's why you're in charge, to control the software process—or, in more politically correct terms, to ensure adherence to the plans you've established, as described in the preceding pages. Therefore, you have to keep checking the plans. And here's an important reminder: you *plan* along the Quadruple Constraint; you *monitor* along the Triple Constraint—the work plan, the schedule, and the project budget.

GETTING AND STAYING IN CONTROL

Today's automated project management systems make it easy to push a couple of buttons and print out a multicolored chart showing how far ahead or behind schedule your project might be, and how you're doing against your budget. However, such reports tell you this information *after the fact*. Your goal should be to avoid problems by knowing about them in advance. If you've properly planned, you've included activities and milestones that allow you to detect problems early. For example, if you have enough levels of testing in the construction management (CM) software package that is being integrated into your corporate accounting system, you might detect internal inaccuracies in the CM system early enough to correct them quickly and easily.

Reporting to the Right Reader

Like any well-written document, a report should be designed for the audience it is intended to serve. And therein lies the rub. Many constituents are interested in a project's outcome. If it's an enterprisewide system, or if future software is to be built upon it, then clearly the project's progress is important to others outside the project.

For example, my team implemented a strategic information systems and data communications plan for a client. The first system was a hardware, software, and network infrastructure system. There were three dozen subsequent projects, all of which depended upon the successful outcome of that initial project. Reporting was necessary not only to the executive committee that sponsored the entire strategic project, but to:

- The entire company, because each desktop was disrupted when equipment was installed.
- The receiving department, because large shipments were coming in from diverse sources and had to be efficiently deployed for installation.
- Security, to monitor the steady stream of subcontractors performing the installation.
- Customers, whose phone and existing computer lines might be disrupted.
- The accounting department, who needed to know where to charge equipment expenses, where to charge back labor, and so on.

I know you get the point.

Granular Reporting: Leveling Detail

Less obvious is that different levels of details are required in the reporting process, depending upon who will read the report. Getting the right level of information to the right people will enable your project to proceed more effectively. The higher up the organization you go, generally, the less detail is required in a project report because your project report is often combined with others. On the other hand, as project leader, you will need *more* detail so that you can quickly identify where a problem is now or might arise in the future. You will need to have enough information at your fingertips to solve it quickly.

The result is a pyramid like that shown in Figure 5-1. This depicts the number of projects being managed at different levels within an organization, and the different levels of detail required to manage at that level. Using today's technology, especially the power of cut and paste, you can put an executive summary paragraph at the beginning of your detailed report, for copying into your department or division management's high-level summary. An example of such a summary paragraph is given in Text Box 5-1. The detail appropriate to the project, however, should follow for each project.

You can also prepare a one-page chart showing an overview of progress at the program level (multiple projects loosely related by common funding, location, and/or sponsorship, and coordinated). Though you will need less detail on each project, you will have to

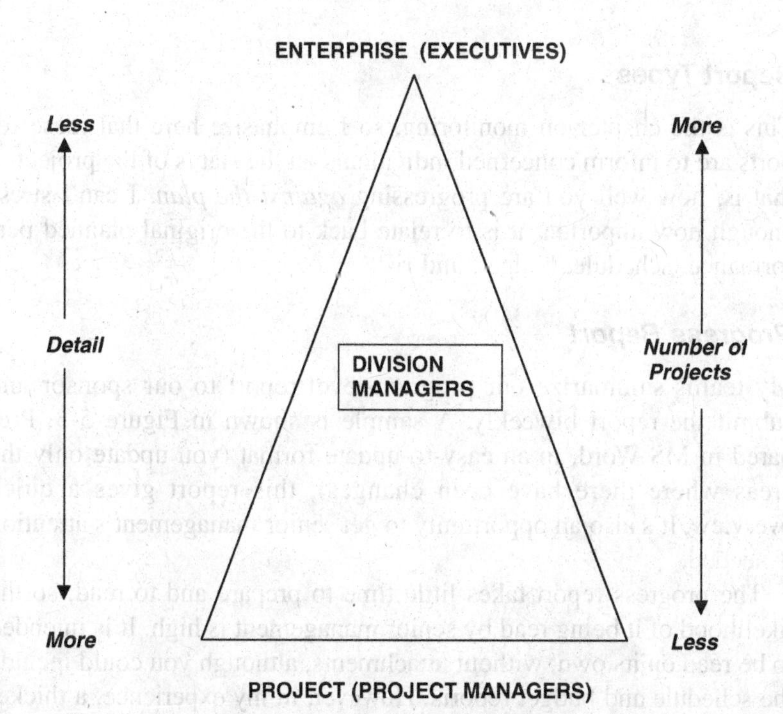

Figure 5-1 *Report detail hierarchy.*

TEXT BOX 5-1 SAMPLE EXECUTIVE SUMMARY

Information Systems Implementation Plan

The implementation of the 25 component technology projects is proceeding according to plan. Fifteen projects are ahead of schedule, five are on schedule, and five are between two and four weeks behind schedule due to delayed hardware deliveries. None of the delayed projects are on the critical path, and no impact is foreseen to overall timely program completion. The program continues to progress within budget, showing 10 percent less expended than forecasted for this period. There are no performance issues, and hardware deliveries are anticipated in the next reporting period.

"drill down" to more detail to provide the reader with more information. See the overall schedule in Figure 5-2.

Report Types

This is the chapter on monitoring, so I emphasize here that these reports are to inform concerned individuals on the status of the project—that is, how well you are progressing *against the plan.* I can't stress enough how important it is to relate back to the original planned performance, schedule, budget, and risk.

Progress Report

My teams summarize our program-level report to our sponsor and submit the report biweekly. A sample is shown in Figure 5-3. Prepared in MS Word, in an easy-to-update format (you update only the areas where there have been changes), this report gives a quick overview. It's also an opportunity to get senior management's attention if needed.

The progress report takes little time to prepare and to read, so the likelihood of it being read by senior management is high. It is intended to be read on its own, without attachments, although you could include the schedule and budget reports. However, in my experience, a thicker report is not read as promptly.

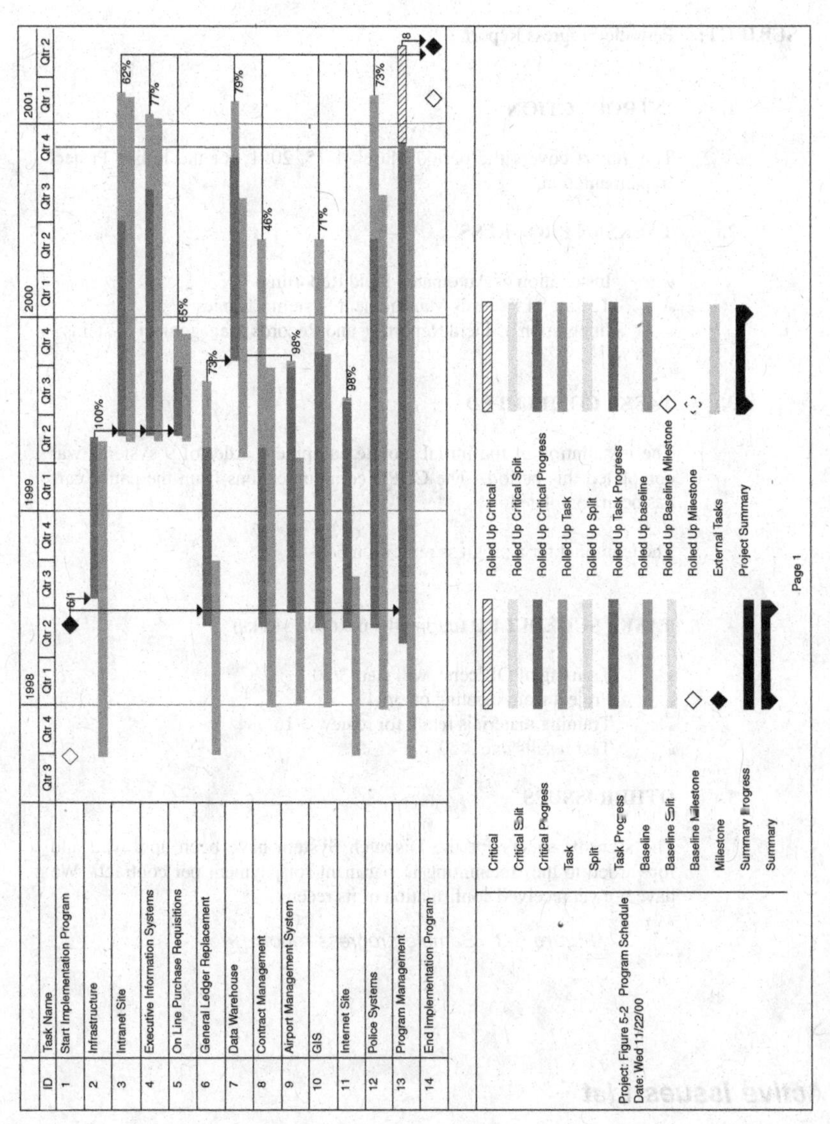

Figure 5-2 *Program schedule.*

TO: Program Manager

FROM: Project Manager

SUBJECT: Periodic Progress Report

1. **INTRODUCTION**

This report covers the period March 1-15, 2001, for the Police Project Implementation.

2. **TASKS IN PROGRESS**

- Installation of Automated Field Reporting
- Testing of Records Management System Changes
- Integration of Field Reporting and Records Management Systems

3. **TASKS COMPLETED**

The installation of the initial mobile equipment order of 9 systems was completed this period. The CDPD communications from the patrol cars has been completed.

The team met for a progress review on 3/14.

4. **TASKS SCHEDULED** (during the next two weeks)

- Training of Officers: will start 3/20
- Project team meeting on 3/24
- Training materials ready for review 3/16
- Test results due 3/20

5. **OTHER ISSUES**

The sign-off sheets for the Dispatch System have been approved and forwarded to the Accounting Department for payment per contract. We have not yet received confirmation of its receipt.

Figure 5-3 *Sample progress report.*

Active Issues List

From project inspection, there will be design and procedural issues that require making one or more decisions. Often, you're into the design phase, and even into development, before enough information has been gathered to make these decisions. A danger is that with so many activities going on, such issues may be overlooked or decisions may not be made in a timely fashion.

To prevent the oversight, we keep a simple Excel spreadsheet list of active issues, as shown in Table 5-1. It has detail on the issue—such as when it was first raised and who is responsible for responding—and summarizes progress to date. The thread of discussion shown in the rightmost column is important, in that anyone reading the report can see the evolution of the issue. Due dates are also specified so that the decisions can be made when needed.

It is common for many issues to arise and exist during the initial project stages. We typically encounter issues related to:

Procedures. How will information in the new system be acquired, supported, and communicated to others?

Database conversions. How will different field sizes be handled, and new fields be derived?

New features. How will features not currently available be implemented? How will current features be implemented using the new system? What gaps will exist?

Interfaces. What data formats and other handoff criteria must be satisfied with the new software?

Technology. Are middleware products needed? What network infrastructure is required?

Authorization. Have the necessary specification document approvals been received?

As the project moves ahead, items move off the active issues list to a closed issues list, and are archived as part of the project records. At the end of the project, all issues should be on the closed issues list. I eyeball the open issues list frequently during the project—its length quickly indicates to me the progress being made on the performance aspect of the Quadruple Constraint. A thick report indicates there are too many unresolved issues and that the project might be heading for trouble.

This report is particularly useful for the development team and for contractors on your team. There is enough detail so that the issues can be understood, and addressed individually or in a meeting.

TABLE 5-1 Sample Issues List

Last Update: 07/10/2000 **Legend:** C = customer, V = vendor, B = both

#	Date Open	Description	Assign To	Priority	Date Due	Date Comp	Status/Comments/Disposition
13	12/15	Send validation tables to customer for verification.	V PM	2	~~5/31~~ **8/15**		customer 011100: When might these be expected? Vendor asked for changes to be penciled in to make identification of them easier. customer 061500: The Startup Guide (with base validation tables) was received. Field mapping to Dispatch fields needs to be completed before we can verify that the validation tables are correct.
27	2/2	Make changes to Dispatch Functional Spec Document discussed during 2/2 meeting.	V	2	~~5/5~~ ~~6/30~~ **8/30**		customer 020200: This should include the addition of Automated Field Reporting (AFR). customer 022800: See 2/2 site meeting notes #5, #6, #7, and #8 for more details.
53	2/2	Develop and obtain approval of Automated Field Reporting Functional Spec Document.	V	2	~~4/28~~ ~~6/30~~ **9/15**		customer 022800: Issue originates from Jan. 11 FSD Review #2.2.4, #3.1.13, #4.1.4, and #4.2.2. See 2/2 site meeting notes #10 for more details. Active Issue Item #12 was closed and "rolled into" this item. customer 050300: Need to update due date. Provide date by 5/12. vendor 062000: This item cannot be completed until #92 is completed. vendor 070900: Forms are still being reviewed (#92 is not complete yet).
92	5/15	Add subcontractor sections to SOW.	V	3	~~6/30~~ **8/15**		vendor 051500: This item originates from closed Item #48.
93	6/2	Configure and ship Dispatch tutorial machine.	C	2	6/30	6/25	vendor 060200: This item originates from closed Item #15. vendor 062300: Machine was shipped on 6/23. Customer 6/25: Machine received.

Bean Count

I've worked on many projects where project status was summarized to a count of issues, late activities, or modules completed. For example, a project summary might show:

	This Period	Last Period
Open issues	25	22
Modules completed	3	5
Testing completed	2	3
Bugs discovered	25	10

This quick look at a project is an attempt to show a trend, but it doesn't give a true picture of the project, for two reasons:

- Without reference to the planned numbers for the two periods, we can't tell whether the performance was better or worse than anticipated.
- Since not all modules, tests, or bugs are equal in complexity and importance, the categories tell us little in terms of whether we should be relieved or worried with the information.

We could add another column, such as Planned or Forecasted, to indicate the monthly trends, then graph them (which is always more helpful to communicate a trend). However, the more we add to bean counts, which are intended for a quick indication, the more time they take to comprehend, thus defeating the purpose.

Bean counts remind me of those mock thermometers used during fund-raising campaigns for charities: as the percentage of completion (100 percent participation) rises, the level of paint—usually red—in the thermometer goes higher and higher. What it doesn't indicate is if the amount of money being contributed is meeting expectations. A better indicator would be percentage of financial goal, say $10 million, that has been achieved.

Bean counts can, of course, be used to motivate the team, but the problem still exists that, for example, having a lower-than-anticipated number of detected bugs does not mean you're doing well. It may mean you just haven't found all the bugs yet and that you and the project are in for a few surprises! In short, I don't recommend the use of bean counts for software projects.

Verbal Debrief

There is no substitute for sitting down and talking with your sponsors, with your team, and with your vendors or contractors. In this e-everything era, don't forgo entirely the value of oral presentations. They can help to:

- Deliver bad news with a positive spin.
- Answer questions that arise.
- Maintain team enthusiasm.

Rather than wait for a milestone review, as described shortly, I try to have informal meetings with sponsors to keep them updated on progress. If there is an issue, such as an alternative design implementation or a staffing alternative with departmental impact, an informal meeting allows for freer discussion and timely decision making. And to address sensitive issues, such as personnel problems, you may want to hold an informal verbal management debriefing without benefit of written materials.

Report Frequency

Reports should not be prepared more frequently than you have meaningful information to report. As mentioned earlier, if your accounting system issues checks on a semimonthly basis, a weekly report will not contain enough new information to be worth the reader's time. Since schedules generally have activities covering a two-week period, biweekly is usually often enough.

However, on very short projects, of, say, eight weeks' duration, you'll be halfway through the project on your second biweekly progress report, making any remedial action harder to take without affecting schedule or budget. So a second criterion for reporting is never report less frequently than you can take action upon the findings. On projects lasting two months or less, report weekly if you can, on schedule; otherwise, reporting every two weeks on all project components is adequate.

Rolling Up Multiple Projects

Often, you'll find yourself managing a project that is part of another project. For example, in our program-level reporting, we would want

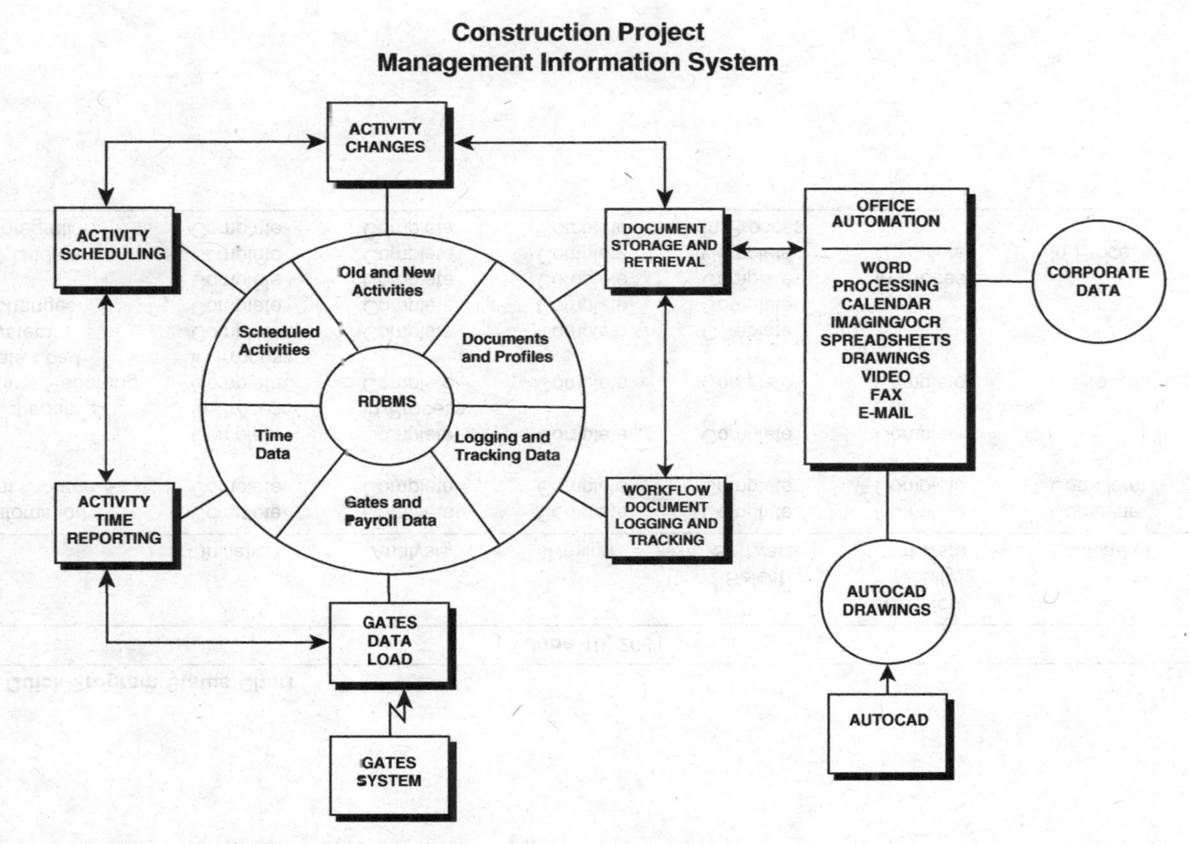

Figure 5-4 Program architecture chart for multiple projects.

TABLE 5-2 Quick Program Status Chart

			June 10, 2001				
Project	Initiation	Analysis	Design	Select Software	Develop/ Modify Software	Implement	Review
1. Office Automation	Complete	Complete	Complete	Complete	Complete	Complete	In Process
2. Document Storage & Retrieval	Complete	Complete	Complete	Complete	Complete	Complete	Complete
3. RDBMS	Complete	Complete	Complete	Complete	Complete		
4. Activity Scheduling	In Process	In Process					
5. Activity Time Reporting	Complete	Complete	Complete	Complete	Complete	In Process	
6. Gates Data Load	In Process						
7. Gates System	Complete	Complete	Complete	Complete			
8. Activity Changes	Complete	Complete	Complete	Complete			
9. Workflow	Complete	Complete	Complete	Complete	In Process		
10. AutoCAD Update	Complete	Complete	Complete	Complete	Complete	In Process	
11. System Integration	Complete	Complete	Complete	In Process			

Gray = complete

to include not only how any individual project is progressing, but how the program as a whole is progressing. That means rolling up all projects into a single schedule, consolidating the budgets for all projects, and, typically, including a systems chart to show the interrelationships of the projects. See the architecture diagram in Figure 5-4.

A quick way to represent progress on a program level is to color-code the systems interrelationship diagram: use different colors for the three states—complete, unstarted, and in process. This gives a quick, high-level understanding of the overall project status. However, as noted earlier, this is an overview for management or introductory purposes; the detail should be supplied in the project-level reporting. A sample of such a chart is given in Table 5-2.

FINANCIALLY SPEAKING

Since project costs typically relate closely to the companywide accounting practices and systems, I've found the hardest part of monitoring project budgets is getting the necessary information from the finance and other departments in a timely manner. For example, in one company, the purchase requisitions for hardware, software, and installation services seemed to disappear into the ether; no copies were available of the final purchase orders, receipts of delivery, and invoices for the installations. For us, trying to keep accurate records of how much had been expended against budget was like working with a ouija board.

Because enterprisewide accounting departments have to be concerned with issuing audited data, they need additional time to verify against internal checks and balances that monies to be paid were indeed paid. By the time you learn you've exceeded the budget, it's too late to do anything about it. So to extend a project budget report's usefulness, you can add a column for an encumbrance, or obligated, amount. There are three columns for each expenditure type:

Budgeted Obligated Paid

In project work, it's best to work with the obligated amounts: this reflects the amount you've already requisitioned or contracted for, whether or not the vendor has been paid. Table 5-3 presents a sample encumbrance report. When the invoices have been submitted and paid, you move the amount to the paid column and reduce the obligation.

TABLE 5-3 Sample Encumbrance Report

Sample Encumbrance Report

Project: Police Management System

Vendor: 64532

Date	Budget	Obligated	Paid	Document No.
06/01/00	$250,000			
01/04/01		$25,000		P.O. 23465
01/20/01			$25,000	check 2766
01/07/01		$14,000		P.O. 23468
Total	$250,000	$39,000	$25,000	

Unused Amount: $250,000 − $39,000 = $211,000
Cash Available: $250,000 − $25,000 = $225,000

Recently my team worked with a client whose project management staff was close to rebelling: they couldn't tell if the hardware order had been properly placed, or identify the final cost of the software ordered. That caused a delay in making subsequent purchases, because while they didn't want to exceed budget, they needed to get as much hardware and application software as possible. As a remedy, we suggested that no purchase order be placed or payment be made until someone on staff had personally signed the order and the pay request. Though this, too, introduced delays in the system, it allowed the project manager to log in to his or her own cost control system (in this case, a simple spreadsheet) the obligated and paid amounts. At any point, the spreadsheet was more accurate than the accounting records.

Determining Project Status

To determine how your project is performing on the budget dimension, you must know how far along you truly are. In the past, percent complete was used as the defining factor; if, say, you'd expended 80 percent of your total project budget, then you were 80 percent complete. But if you were only 20 percent into your project schedule, did that mean you were 60 percent ahead of schedule? Unfortunately, it meant only that you were 20 percent into the project, and had spent 80 percent of the budget. Unless you had additional project management information, such as how the resources were planned to be expended over time—say, a cash flow—you didn't know if you were where you ought to be.

I recall on one assignment that the expenditures were well above where the project was in schedule. The project manager observed that the project had been heavily "front-loaded," meaning that more monies were anticipated to be spent early in the project. But in software projects, the total budget amount can easily be exceeded if the developed software, upon testing, turns out to require extensive reworking. So frontloading in software projects may not be of any help.

Taking the Earned Value Approach

This problem of you're-not-really-complete-until-it's-over led to the implementation of *earned value approaches* to cost and percent complete. Earned value compares the amount of planned work with that actually accomplished, to determine whether cost and schedule performance are on target. However, since it does not address the *quality* of what has been developed, you may have spent both time and money to stay in the same place. Nevertheless, earned value analysis is the most commonly used method of measuring performance.[1] It is a percentage of the total budget equal to the percentage of the work actually completed. This figure is then compared with the anticipated schedule and budget to determine variances from the plan.

I think the entire area of earned value for software development is overkill, due to the uniqueness of each software project, which makes it hard to determine what value in your situation 700 lines of untested code might produce. I don't recommend it for software projects. Instead, the project manager should emphasize testing at early project stages.

Tracking Expenditures

A difficulty in cost reports is ensuring the accuracy of charges made to your project. Emphasis should be placed on those financial charges that are well beyond your control but that affect your project budget. Remember the G&A and overhead rates explained in Chapter 3? Periodically, the rates may change, and your project charges will reflect them accordingly. There is nothing you can do other than try to compensate by saving monies elsewhere.

[1] *A Guide to the Project Management Body of Knowledge* (*PMBOK® Guide*), 2000 edition (Project Management Institute, Newtown Square, PA), p. 123.

Figure 5-5 shows an estimated expenditure curve over time, with a line specifying the actual expenditures incurred. From this, you can readily determine if you're above the planned curve, hence spending more than you should. But what this representation doesn't tell you is if you're just racing through the project, when, in fact, you should be spending more because you're accomplishing more. That might be the case if your programmers are performing above expectations, and as a result you've moved further into testing and implementation than originally scheduled.

What do you do when you find your project is ahead on spending but not on schedule? The distance above the planned expenditure line gives you a hint of the trend that's emerging: at 15 percent, perhaps it's time to rethink where the project is, and come up with a plan to bring the costs back in line with the schedule, so that you don't end up "throwing bad money after good."

Presuming you need additional funding, how do you get it? That's where the contingency helps. By having that 15 percent contingency I alluded to earlier, you have some funds to use to bring the project back into line. Of course, if the project requires still more funding, perhaps for additional programming or new platforms, then you'll have to ask your sponsors for more monies—in which case you'd better be ready to fully explain why they should grant you more funds to accomplish the same goal.

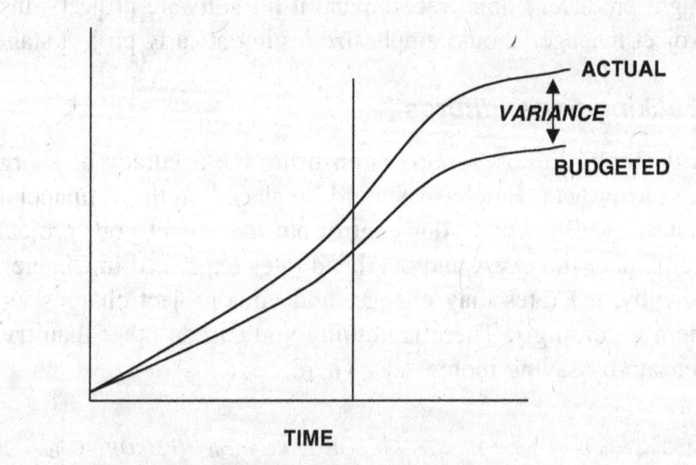

Figure 5-5 *Planned versus actual costs curve.*

Conducting Periodic Reviews

When a project gets into trouble, it's time for a review, of course. But there are other reviews you should plan during the life of your project. The need for written reports does not obviate the need for periodic human interchange in a project meeting. These interim meetings during the life of the project are separate from the postproject or project completion reviews, covered in the next section. The primary purpose of these reviews is to discuss progress and keep your team motivated.

Whatever the primary purpose of the review, you should make sure that you have created an agenda and distributed it to the attendees; and don't forget to send copies to their managers if the political situation in your company warrants it. Prepare for the meeting by creating handouts of critical information; and use a Proxima-type projector to display a PowerPoint-type presentation to the attendees. If you are unsure of how to prepare a presentation, look for a wizard in your presentation program. Usually, many helpful templates are provided. For example, Text Box 5-2 shows a presentation agenda for a review.

At the outset of each review, restate the purpose of the review: decision making or progress reporting. If you have to do both in a single meeting—which you should avoid whenever possible—separate the actions into two segments (typically, follow progress reporting with decision making). That way you'll be able to make specific decisions after the attendees know where they are and what the significant issues are. For example, if you have an architectural decision to make regarding which platforms to implement, first determine where the entire project is, lest you decide something that will make the project more precarious.

Reviews are meetings with specific purposes, so following these rules for productive meetings will stand you in good stead:

- Stay on the subject.
- Stay on schedule.
- Avoid letting discussions become personal.
- Identify and reinforce the critical issues.
- Assign knowledgeable people to work on identified problems *outside of the meeting* and to report back by a specified date.
- Assign every action item a due date and a single responsible individual, although you may appoint others to assist.

TEXT BOX 5-2 SAMPLE REVIEW MEETING AGENDA

1 Police Project Progress Review
   Company, Project, and Project Manager Names

2 Status Summary
   - *Is project on track for implementation as expected?*
   - *Identify outstanding items needing resolution.*

3 Progress
   - *List achievements and progress since last status update was given.*
   - *Review last meeting's action items due for completion.*
   - *List delays, problems, and corrective actions since last status update was given.*

4 Schedule
   - *List top high-level dates.*
   - *Distribute more detailed schedule if appropriate.*

5 Implementation
   - *List main critical deliverables.*
   - *Identify test progress and results.*

6 Costs
   - *List new projections of costs.*
   - *Identify change orders: amounts and reasons.*
   - *Identify future cost trends.*

7 Technology
   - *List technical problems that have been solved.*
   - *List outstanding technical issues and impact.*
   - *Identify any dubious technological dependencies for project and backup plans.*

8 Resources
   - *Summarize project resources.*
   - *Identify any shortfalls or surplus.*

9 Goals for Next Review
   - *Assign date of next status update.*
   - *List goals for next review.*
   - *Summarize action items.*

- Distribute critical documents prior to the meeting, so that attendees can come prepared to contribute and participate.
- Don't dictate "from the chair." If you have strong feelings, hand the chair to someone else when you express them.
- Appoint a scribe to take notes during the meeting; and instruct that person to distribute the notes promptly thereafter to attendees and identified third parties.
- Reserve in advance whatever equipment you require for the meeting (projectors, computers), as well as the conference room, so that attendees spend time working, not wandering around looking for a place to work.

Milestone Reviews

Milestone reviews should be planned into your WBS, and included on your project schedule at the end of a project phase. Depending upon the project life cycle in your company, your phases may be different from those shown in Figure 2-4 earlier. Since a specific document or set of documents is produced at the end of each phase, an appropriate agenda item is to review the content of the document set.

Each review may include different individuals who are appropriate to the milestone under review. For example, at the end of the requirements phase, your users should be heavily represented. At the end of the implementation, you should have user participation, along with your developers and/or vendors. The project team typically comprises the primary attendance, with other groups and departments represented, such as quality assurance and network services. Keep in mind, usually it's better to invite more people than less, so that you have a wide range of views in an appropriate forum.

How does the project team relate to the design review? Typically, it is a subset of the attendees: all users may attend, though only a few are from the original team.

The format of the review, however, should be consistent, and should aim to:

- Identify any issues requiring resolution, prior to moving on to the next stage.
- Gain consensus.
- Share lessons learned that might be helpful for future activities.

Management Reviews

Like reports, reviews need to be granular, with different levels of management represented, especially executive levels. Usually, the executives are the ultimate sponsors of your project, although they may have little time for frequent interaction during the course of a project. Reporting along the Quadruple Constraint of cost, schedule, performance, and risk, at a high level, will make these reviews meaningful to the executives, and ultimately to your project team.

Management meetings may be more difficult to arrange due to busier schedules at that level in the organization, but they are important for maintaining visibility in the organization. Often, I've learned about political changes in the offing that will affect both sponsorship of the project (managerial issues) and processes involving the project (technical and interface issues). Hearing about them early enough is the only way to adequately plan for them.

Who attends these meetings is generally determined by the executives, but usually they are limited to the project manager and, possibly, the lead architect or designer. Shorter meetings are better than lengthier ones, for the simple reason that they are easier to squeeze into the busy executive schedule.

> *Note:*
> Don't confuse management reviews with technical reviews, which are called for in many design methodologies, such as a structured walkthrough. In the latter, programmers read and critique one another's code. This is part of the design process, not the project management process. While the technical review should be an activity in the WBS and on the project schedule, it is not a project management review.

DEALING WITH CREEPY CRAWLY CHANGES

Lewin's Law of Random Perversity states that there will always be more changes in a project than you want.[2] You cannot avoid them, so

[2] Lewin and Rosenau, *Software Project Management: Step by Step,* 2nd edition (Marsha D. Lewin Associates, Inc., Los Angeles, CA, 1988), p. 238.

you must learn how best to deal with them. Recently, a client of ours awarded a contract to a highly qualified vendor in response to a complete proposal. Later, changes in telecommunications technology and new software releases required altering the original proposal. There was no way these changes could have been foreseen, because:

- The length of the contract was longer than the vendor's software release cycle; the content of the release was not known at contract-signing time.
- The telecommunications capabilities themselves had advanced over time.
- No one had prior experience in marrying the hardware, software, and telecommunications.

In this case, fortuitously, the increases in costs were offset by decreases in components originally budgeted for and no longer needed. The change order was actually a net credit to the contract.

The really difficult changes to deal with, however, are those that occur because the project was inadequately planned, or because poor estimates were made. That said, even the best estimates today will have some error because the technologies change so quickly. In short, it's better to plan for change rather than resist it. Resistance in this case is, indeed, futile!

Issuing Change Orders

How often should change orders be issued? Typically, change orders can be put through whenever the contracting parties agree. As a practical matter, however, unless the budgetary impact is going to be severe, grouping change orders together is more efficient. It's also better to ask the approving body to authorize changes infrequently, regardless of the actual amount being requested. A single aggregated change order for $100,000 on a $2 million dollar contract (5 percent) costs less to process, and isn't as annoying to the approvers as eight changes of $12,000 each! Don't nickel and dime your executives to death!

Having said that, I recall a consultant who started in the field when I did. We shared the same client, and every time the client asked him to tweak a report or add a feature, he charged the client. I chose to do small changes without increase. He nickeled and dimed the client, and built a flourishing national business for himself and hundreds of employees, while I enjoyed the time I saved writing change orders. You can judge for yourself which style you prefer.

Change orders are also best prepared after the underlying impact of changes has been assessed and fully scoped out; typically, this occurs within a project phase. Subsequent phase cost impacts can also be requested at that time. For example, if you decide to switch to an NT Server instead of a UNIX box, in your change order, you might need to include workstation hardware upgrades for better performance, as well as the cost of the server and additional vendor involvement for testing.

A sample change order is shown in Figure 5-6. Your company may prefer a different format; the key is to have the necessary information on the change order to clearly define any schedule, budget, perfor-

Originator: Vendor Project Manager
Date of request: 3/15/01
Need/return by: 4/1/01

NOTE: This will be Change Order #1 for the client.

Client Name:	Client Company
Client Contact:	Marsha Lewin
Client Address:	255 Main Street Everytown, CA 90025
Description of Change:	**REMOVE** - From Exhibit A, **Software**, remove **Server Queuing Software** ($3,500.00) - From Exhibit A, **Services**, remove **Master File data conversion**. ($7,500.00) - Reduce License and Maintenance fees accordingly. **ADD** - In Exhibit A, under **Services**, add **Master File format customization** ($2000) **CHANGE** - Change Payment Milestones accordingly using credits and/or fully changed amounts.
Change Amount:	-$9,000.00 one-time cost, -$300 Annual fees
Change From (check one)	Original Contract _X_ or Prior Change Order ___ or Other ___
If Change from "Other"	Description of Other:

Figure 5-6 *Sample change order format.*

mance, and risk impact. The change order should be signed by the contracting parties.

Controlling Changes

One of the wonderful aspects of a software project is seeing users involved in the design of the software in a prototyping environment. But they can become so excited about, and pleased with, their own creative participation that they sometimes forget that prototyping is not an eternal process. Throughout the implementation, they consistently suggest, and often feel they *must have,* more changes. My team generally institutes change control when the functional design specification of the software has been agreed to by the users and the development staff or vendor. We call this a *design freeze.* A critical issue with design freezes is that unless you stop making software changes long enough to finish developing the software, you'll never be able to test it fully and complete the project.

You can, of course, also institute change controls when the requirements have been determined, but by doing so, you often end up making more work. Today's packaged software may not satisfy all your requirements precisely, but you can often eliminate some by utilizing other packaged features you hadn't previously considered. When you take this route, you then should draft a document that maps how the requirements were satisfied by the vendor's product. File it away with the other project documents to provide a complete audit trail of changes suggested. An example of such a document map can be found in Table 5-4. Such a document is important should you ever be asked why certain features were not included in the final implementation.

The sources of changes are many: internal, due to process, political, or technology changes, and external, due to technology or changes introduced because of interfacing requirements with others' systems. Some are within your control; most are not.

Implementing Change Control Systems

Every software project should have some method of accommodating requests for change throughout its life. The method should suit the company and the project's scope and duration. A change control system may be thought of as a formal approval cycle for changes to any approved milestone document. That would include program code, operations procedures, and functional design specifications, for example.

TABLE 5-4 Audit Trail of RFP Requirements

RFP Requirement #	RFP Requirement	Original Vendor Response	Final Disposition
7	Provide audit trail, to reflect who entered or inquired against data and when.	Comply. All transactions are audited.	Complete audit trail of transactions is still under development and may not be available prior to scheduled "go live" date. Manual workaround in interim is acceptable.
9	Link images to case, including photos and sketches.	Comply. Vendor has proposed separate imaging system.	Client has document management/imaging system being implemented as another project. Vendor will develop API.
11	Provide version control over documents.	Vendor does not maintain versions of same document.	Client's document management system will handle version control. Vendor will supply interface to that system with on-screen button.
18	Provide nondisclosure reminder notice on each page of reports generated.	Exception. Vendor may be able to comply with this but additional information is required.	Vendor will provide for additional funding (see Change Order #2)
109	Viewing and inquiry—Allow viewing of any report, with embedded images, sketches, pictures, moving video, or voice note.	Not all subsystems have capability.	This is covered in Change Order #2.

The purpose of change control is to ensure that any changes made to schedule, budget, risk, or performance are consistent with the project goals, and worth the additional expense and/or delay incurred as a result. The scope of change control includes determining if the change is beneficial and monitoring the approved change until completion.

At a minimum, users should be instructed to initiate changes using a paper or online change request form, an example of which is shown in Addendum 2 of Appendix I. The form should indicate what benefit the change will provide, such as: increase data entry time by 3 percent; eliminate manual calculation errors on chargeback rates for airport concessions. The more detail the change originator can provide, the easier it will be to scope out the potential change—in terms of cost, length of time it will take, and any reduction in risk to the project. Good change control systems are integrated, meaning they consider the effect it will have on all project dimensions.

In some organizations, the change request is then submitted to the technical staff, to identify the scope of other changes that might snowball from the first. Hardware, software, documentation, and a rough schedule of implementation and cost might be provided. In other organizations, the change is not scoped out more fully unless it is approved by the change control board.

The change request is next submitted to a change control board, which is composed of individuals capable of assessing the justification for the change and approving its functionality. (Cost may have to be approved by a separate budget committee.) During the development phase, the change control board is typically made up of the project team. When the project is complete, generally a user board is set up, spanning all software being supported by the IT department. Any changes subsequent to completion of the project are submitted to that entity for approval through a similar process. Wherever possible, users should be part of the change control process, as in the design process.

The change process flow is shown in Figure 5-7. The change request, in many ways, is a request for a miniature project, and when approved becomes a contract in miniature, specifying the Quadruple Constraint of performance, cost, risk, and schedule. The change process should also include updating all relevant documentation and any training necessary.

Depending upon company size and the level of software support available, changes can be managed with sophisticated computerized systems, a simple spreadsheet, or manually. A report of changes

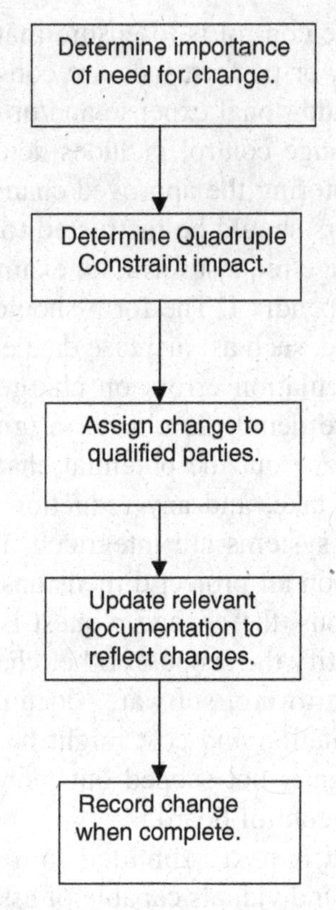

Figure 5-7 *Change process flow.*

proposed and their disposition (approved, denied, or in process) is generally prepared for departmental management, so that everyone knows the disposition of their requests.

If there are significant changes to your project, you may want to implement them after the base software has been delivered so that users gain some benefit from the system. I recall on one system, the primary user was reluctant to leave well enough alone, and kept requesting additional changes long after the system was completed. These changes were of the nature of eliminating a single keystroke or changing the color of a screen or an icon—not significant enough to warrant the cost. Unfortunately, the sponsor approved such changes without the necessary cost and benefit justifications, resulting in avoidable expenditures of thousands of dollars.

Finally, changes should be made in writing and be well documented. Then, if you ever have a project that overruns its budget or is delivered late, you'll have a record of what occurred, why, when, and on whose authorization. And if the performance has changed from what was originally specified, you'll also have an audit trail.

COMMUNICATING AND PUBLICIZING

A few words about spreading the word about your project. Occasional e-mail to interested parties within your organization, either under your e-signature or your departmental manager's, lets others know what's happening. If your company permits it, sending global e-mails on a well-planned, infrequent basis may also be appropriate. Often, a process will be altered by the implementation of new software. When personnel know about the changes, they can help institute the new processes more efficiently.

Undoubtedly, you will not always have good news to spread—the project is not satisfying its Quadruple Constraint, for example, and there will be obvious overruns. While there is an understandable and natural tendency to hesitate to publicize bad news, the sooner you get the information out there, the better. Delaying only makes the bad news worse. The best course of action is to promptly replan and present it in the most positive way possible.

In closing, let me say that all project managers at one time or another will fall prey to Murphy's Law—everything that can go wrong will go wrong. Typically, this will happen when least convenient. My best advice is to face these challenges with tolerance, understanding, and a sense of humor. You can control changes, but not eliminate them.

6

Completing

It's not over 'til it's over. We all know that. But with technology always changing, vendors offering frequent upgrades, and users always requesting more, it's more difficult than ever to come to closure, as I like to call it, on a software project. Still, to everything there is an end—followed by the birth of new projects.

KNOWING YOU'RE AT THE FINISH LINE

Assuming you have cleft your project into a Waterfall model, with the phases indicated earlier in Figure 2-4 (the Formula-IT life cycle model), recall the initiation phase. You monitored along the Triple Constraint the plans you developed along the Quadruple Constraint. The project plan you penned guided you through the deliverables, processes, and challenges of implementation. Most important, at the end of implementation (phase 6), you were able to measurably deliver what was promised in the planning stages.

You have your satisfactory testing results and quality assurance report. The change control board has convened. Users have their procedural documentation, the system has its technical documentation, and any maintenance materials have been gathered and organized for handoff to the IT maintenance crew, help desk, and/or user department. In short, you feel you have little more to do other than clear out your belongings and move on.

Making Sure Your Sponsor Feels the Same Way

Whether you've managed the project internally or externally, you have a customer or sponsor who must sign off on the project, verifying that you have satisfied the performance requirements. You dealt with schedule and budgetary requirements along the way, so there should be no surprises on those constraints. I generally ask the customers to authorize in writing that they are assuming responsibility for the system, and that it runs satisfactorily, as they had agreed in the design specification stage. If, instead, the sponsor wants additional features that were not part of the specification—to make the system more usable, for example—this is the time to indicate that there's a basis for another project, now that this one is complete.

You'll also work on projects where the customer is really many individuals, who cannot agree that they are satisfied with the project. You can try to reach consensus by suggesting they work with the system, and assure them you will check with them during the review phase. I've found taking this tack often works. When you cannot get everyone to agree no matter what you try to do, partly because of "fiefdom" issues and partly because some folks are never satisfied, as long as you have the acceptance by the team leader and critical user personnel, proceed to project completion.

On occasion, you'll realize the users are simply fearful of taking ownership of the project. They have become accustomed to your management quarterbacking, and feel rudderless. You can ease any qualms by making sure that they understand the future service-level arrangements and that they know whom to contact for future support. This may take additional time, but it's worth the investment, as a dissatisfied customer, internal or external, will not help you win future management roles.

Shaking Hands and Saying Goodbye

By the time your project is coming to a close, many of the team members will probably have started returning to their other tasks. Likewise, your technical staff will probably have dwindled, already at work on their next programming tasks. And if you contracted out the development, your vendor is chomping at the bit to get his or her final payment and move on to other clients.

Saying goodbye to a vendor is easy: the contract states—hopefully in unambiguous terms—the payment criteria for each step through the assignment. The tests have either been satisfied or not. Well, not exactly: typically there are little nits, such as a field size too small or a missing CD backup that's in the mail. But the contract has essentially been satisfied. The contractor wants final payment, and will pressure you for it.

In general, I delay payment approval until all substantive elements have been satisfied. Experience has shown that no matter how well-intentioned they are, contractors move on, new management takes over, and old promises become meaningless unless they appear in writing (even e-written is preferable to nothing at all!). In one mega-installation of infrastructure, not one key person was left in the vendor's employment 18 months later, let alone working on our project. The point is, only after you have all your documentation (typically the last item), and no tasks are left on your open issues list, should you cut the final payment check—but then do so promptly.

Saying goodbye to your own team, however, is different. This is a wonderful time to write letters of recommendation that can be routed to members' line management, and ultimately become part of their permanent personnel record. (In fact, letters of recommendations for exceptional vendor personnel serve a similar purpose within their organization.) A letter of recommendation is also in order for ancillary personnel who performed exceptionally on your project, such as the purchasing agent, or the receiving clerks who repeatedly gave up their lunch hours to bring over hardware.

Archiving Project Documents

Make sure that records you will need for subsequent project financial audits are archived appropriately according to your organization's policies. Archive project documents in both hard and electronic versions. Today, using CD technology requires the fewest disks. You can also scan hard copies and archive them on CDs. It's best to archive final approved copies of electronic documents in accordance with your organization's records management policy to scan documents. More widespread use of document management systems will facilitate the archiving of proj-ect and other documents in the future.

All meeting minutes become part of the project documents, as do all the milestone deliverables you've amassed during the project. These include:

- Work plan
- Design specifications
- Progress reports
- Test plans
- Test results
- Meeting minutes
- Updates to any of the plans
- Significant memoranda
- Additional reports that may be of future interest during maintenance
- License and warranty information

You may also want to prepare a handoff document that delineates any information that you feel would be of value to the maintenance and user staff.

You can see that a single project's archives may take up considerable space. On a larger project, the handoff documents can fill many boxes. Therefore, you should also provide a packing list, or contents sheet, with the documents. (The final destination of these documents should have been determined during the project planning process. The disposition of any hardware or software procured for the project, but not needed anymore, should also have been decided early on.)

Running the Last .2 of a Mile

As any marathon runner will tell you, at the twenty-sixth mile, there is still .2 of a mile to go to complete the race, and that .2 of a mile is always the hardest! On a software project—which often feels like a marathon—you may well find that getting the attention of your team and any resources to close off your project may be difficult, if not impossible. However, you've persevered this long, so hang on for that last .2 of a mile.

Learning Lessons

As I've said many times throughout this book, every project is unique, but there are also many similarities that you can learn from. I like to hold a lessons-learned meeting at the end of the project of all the involved parties. Everyone is solicited to bring comments, positive or negative, that can be recorded and used as guidance for future projects. No holds are barred; this meeting is not for the faint of heart, and defensiveness is not allowed.

You may want to have someone else chair the meeting, so that you can participate. Remember, it's better not to chair a meeting when you have strong views to promulgate.

Afterward, minutes of the meeting should be distributed as widely as possible, especially to future project management and development staff. In fact, there's generally something of value to come out of these meetings for anyone connected with similar projects, including users. You may find comments, such as the following, of great help in the future:

- The vendor was nonresponsive.
- The software was too buggy to be deployed.
- The hardware supplier provided valuable free consulting.
- The electric outlets at the branch office were inadequate for computer use.
- Fiber optic cable took too long to install.
- Server components were not available promptly.
- We couldn't find an Oracle programmer.
- The new system changed the way we communicate with other departments.
- We need to change our manual processes to get the most out of the new system.

THE JOY OF REVIEWING

The final project phase is review. The purpose is to visit the stakeholders, from users of the system to executive management, to find out how the project has fared since completion. Informal meetings generally work well for this part of the project. Any project goals and lessons-learned documents can also form the basis of questions to ask when you meet with the stakeholders.

The result of this final phase, which generally occurs one to three months after the project has been handed off, is to see how well the project has "taken" in the organization. Have any problems cropped up that aren't being dealt with? Does line management provide the support and resources that every new system requires? Have any unforeseen consequences arisen?

Preparing a brief report for the project sponsor can often reveal:

- Opportunities for additional uses and time savings.
- Need for a full-time data administrator.
- Need for user groups to help share knowledge and provide tips on system use.
- Trainer has left the company and no one has been appointed to assume the training responsibilities.
- The vendor takes too long to answer questions, so the system cannot be used all the time.
- Additional software has been put on the server, and the system seems to run more slowly now.
- Additional licenses are needed, as people often are not able to log on because all available licenses are being used.
- The vendor has issued a new software release providing better reporting that would help the department reconcile its payroll entries.

The fact that you, during your project management task, opened communication paths and forged relationships with others in different parts of the organization will enable you to suggest solutions to such problems uncovered during the review meeting. In some cases, this may mean identifying the need for subsequent projects.

That said, the objective of the review is not to *solve* all problems presented, but to *uncover* them. The change control board can take your suggestions under consideration, as appropriate.

LIFE AFTER COMPLETION

As a result of the implementation, you may find that the organization of the company or departments changes. This is quite common. Systems change the way information flows in an organization. You may be tapped to be a permanent part of the new organization, as a result of your successful project accomplishment and intimate knowledge of the systems! More typically, your mission, should you decide to accept it, will be to fill up your desk with new binders and reports for new projects.

IN CONCLUSION

I'd like to summarize here what I think is most important to know about managing projects. First, most problems you'll encounter will be people issues, not technical issues—users may be unreasonable, the team may exhibit dissension, and you'll become exasperated. Be prepared to deal with these so-called soft issues that can make or break your project.

My advice is, never shoot from the hip. Despite the fact that your role will be to make quick decisions at various points in the project, you will be doing so with the ammunition of much planning and foresight. While any decision you make along the way may not be what had been originally planned, your careful planning and project monitoring throughout will keep you on target to the best course of action.

Here are some other guidelines I think you'll find useful in your career as a project manager:

- *Keep your eyes on the objectives and your feet firmly on the ground.* On many projects, you'll witness moments of sheer panic in one or more parties due to job insecurity, a bad hair day, or any of myriad reasons. These are mere bumps in the road to successful completion, and no matter how hard it may seem, as the project manager, you're responsible for leading your "bad camper" to safer grounds.
- *Learn to overcome the inevitable user resistance.* Listen to the concerns, quantify them, then respond. You'll be surprised how users can be encouraged to come around. Most people really do want to do a good job. When faced with a new situation in which they feel unsure, they worry they will not be able to do a good job. You can ease that concern with your attention and interest.
- *Hold hands.* Stakeholders often have difficulty seeing the final goal, not so much out of obstinacy but out of ignorance of the software development process. Take the time to explain what will be happening before and during the process. Doing so can ensure your project's success. Failure to make that time investment will almost certainly cause delays, if not project failure.

- *Beat the drum.* Projects with high profiles get resources more readily. Create a sense of positive importance and urgency for your project. People want to participate in successes.
- *Recognize and deal with ambiguity.* No matter how hard you try, at some point, there will be a lack of clarity—in the contract payment arrangements, in how the program will interface with the WAN, and so on. You'll have to spend additional hours clearing things up. Don't take this as a personal failure. Management is not a black-and-white process. Your skill is in being able to surf the gray areas successfully. Software projects are, by their nature, more ambiguous than other types of projects. You can reach a successful outcome by applying rigor to the management process, to reduce ambiguity as much as possible, while allowing for the creative juices to flow.
- *Quantify decisions.* Wherever possible, establish objective measurable criteria. For example, assign a numerical value to vendor proposal features, to training, to online help, and to other documentation. This reduces the subjectivity that can cause a project to go awry when parties are not held accountable. It also precludes value judgments, which cause decisions to be taken personally, and often destroy the team spirit so necessary to successful project completion.
- *Put everything in writing.* File e-mails on critical subjects. Confirm critical phone calls with a follow-up e-mail or memorandum to decrease chances of errors due to misinterpretation or omission. Keeping a document trail doesn't eliminate problems, but it can help to minimize them, and it will assist in resolving the problems that do arise.
- *Renew yourself.* You may think you're ready for the next adventure, but take time to make sure you are. Particularly on long-term projects, I've discovered that I become insulated from the rest of the technological world. Go "surf" the industry and see what's new. Especially in technology, today's LAN is tomorrow's server. It's best to keep abreast of your ever-changing tool set!

And there you have it. Another project, another challenge. So it goes. Hopefully, you've learned some tricks of the trade from this book, so that, regardless of the specific technology you work with, you will be better able to manage the process effectively and successfully in the future. Live long and manage well!

Appendix I

Project Plan

Project Plan

This appendix is intended to provide an example of information to include in a software project plan. Items in italics are variables for you to consider and complete.

1. INTRODUCTION

This document describes the plan for implementing a Contract Management System (CMS) for a Sample Corporation (Company). It sets out the following for the project:

- Scope
- Project budget and schedule
- Organization and staffing
- Control
- Management
- Job descriptions
- Task descriptions
- Project standards

An overall project schedule, prepared using MS Project, is provided. A form for requesting a change to a project deliverable is included, as are task descriptions.

2. PROJECT SCOPE

Include here any limitations on the project, and an assessment of risk.

The scope of the CMS project encompasses the implementation of a common system, to provide document and cost control for all Company projects. Particularly, this would involve the selection of a package, or suite of packages, that best satisfies the Company's needs to monitor its construction and design projects.

This project's risk has been assessed as low, and the impact of a failure is assessed as inconvenient. Manual procedures can be utilized on an interim basis if necessary.

3. PROJECT BUDGET AND SCHEDULE

The preliminary cost estimate for this project is $250,000. This comprises:

Software	$100,000
Installation	$ 20,000
Development/configuration	$ 65,000
Training	$ 15,000
Hardware	$ 50,000

Software maintenance, estimated at $15,000 per annum, is not included in the project budget. It is noted only for future budgeting purposes.

The schedule for completion of this project may be found in Figure 1-4.

4. PROJECT ORGANIZATION AND STAFFING

4.1. Project Organization

The CMS project will encompass seven distinct stages:

1. Initiation
2. Analysis
3. Design
4. Software selection

5. Software development/modification
6. Implementation
7. Review

Each of the stages is further broken down into tasks. Each stage has one or more associated deliverable. Detailed descriptions of the tasks are as shown in Text Box 3-1.

1. Initiation
   - Develop Project Plan, including resource requirements and job descriptions.
   - Define project standards.
2. Analysis
   - Analyze requirements.
3. Design
   - Define data, processing, and technology.
   - Define user interface.
   - Compile system design.
4. Software selection
   - Identify list of potential vendors.
   - Develop Request for Proposal (RFP).
   - Evaluate responses.
   - Conduct demonstrations.
   - Select vendor.
5. Software Development/Modification
   - Prepare for development.
   - Build/modify database and processing.
6. Implementation
   - Plan system introduction.
   - Assemble technology products.
   - Instruct trainers to conduct end-user training.
   - Conduct testing.
   - Migrate data.
   - Run parallel/pilot.
   - Initiate operations.
7. Review
   - Review system.

4.2. Project Staffing

The project participants will be:

- Executive Technology Steering Committee Members
- Program Manager
- Program Office
- User Project Teams
- IS team

Figure I.1 illustrates the organization of the project participants. *This is shown from the project perspective.*

The responsibilities are as shown in Figure I.2, Project Roles and Relationships. *This shows the management hierarchy that affects the project.*

The Steering Committee has the responsibility to approve the project deliverables. Once approved by the Steering Committee, project deliverables will come under change control procedures. The Steering Committee is also responsible for periodic review of project through progress reports from the Program Manager.

The CMS Project Team is responsible for completing the project tasks, resulting in the production of the project deliverables within the time frames indicated in the project schedule. Production of the project deliverables will involve a high level of coordination between the Project Team and Company staff.

Detailed job descriptions for the roles of the CMS project participants can be found in Addendum 1 of this document.

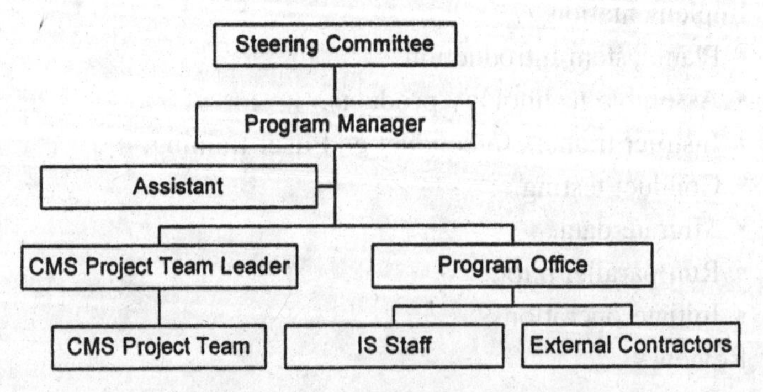

Figure I-1 *Project Organization Chart.*

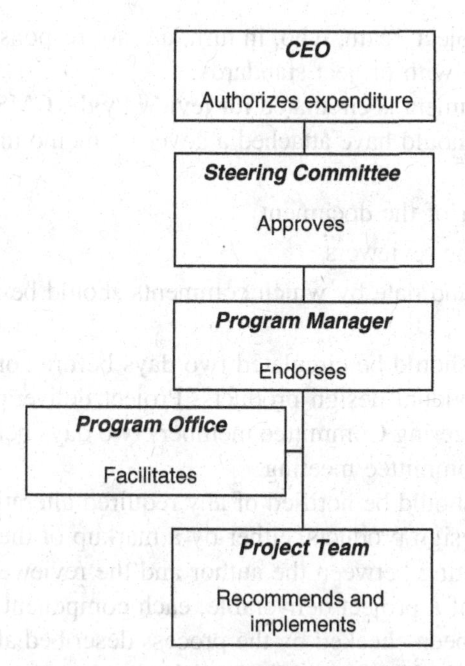

Figure I-2 *Project Roles and Relationships.*

5. MANAGEMENT CONTROL

Define who is responsible for writing and approving documents, from both content and standards conformance perspectives.

Responsibility for management control of the project ultimately rests with the Program Steering Committee. This responsibility is exercised through periodic (approximately bimonthly) review of progress, and approval of project deliverables. This acceptance of project deliverables by the Steering Committee is deemed to be acceptance by the Company.

Each completed CMS project deliverable will be given to the Program Manager for approval. The Program Manager will check compliance with project standards, including adherence to the task description and style. The Program Manager will also arrange for any review by the appropriate parties, as necessary.

For interim products (as opposed to project deliverables), this will generally involve approval by the CMS Project Team, with the assistance of the Program Office. The Program Manager will not be responsible for checking the quality of content; this is the responsibility of the CMS Project Team, who, in turn, are not responsible for checking compliance with project standards.

When a document is circulated for review by the CMS Project Team, the document should have attached a covering memo that includes:

- The author of the document
- A list of the reviewers
- The time and date by which comments should be received

Documents should be circulated two days before comments are required for individual design products. Project deliverables should be circulated to Steering Committee members two days before the date of the Steering Committee meeting.

The author should be notified of any required alterations or amendments to the design products, either by a markup of the original product or at a meeting between the author and the reviewer.

In the case of a project deliverable, each component of the deliverable will have been checked by the process described above. The final deliverable need, therefore, only be submitted to the Steering Committee for approval without further quality assurance. The Program Manager will review the integration of the components for consistency and adherence to project standards before submission.

Where changes to a deliverable are requested by the Steering Committee, the nature and form of the changes will be agreed to at the Steering Committee meeting. The author will then make changes to the document, and the document can be signed off at the sole discretion of the Program Manager.

In this manner, each design product will receive four levels of review:

- By the Project Manager, to ensure compliance with task description and project standards
- By user representatives, as an individual design product
- By the Steering Committee, as part of a project deliverable
- By the Steering Committee, as a part of the final design report

Once a project deliverable has been accepted by the Steering Committee, it will come under formal change control, as described by the change control procedures in section 6.5.2, Document Change Control.

6. PROJECT MANAGEMENT

The CMS Project Team Leader is responsible for overseeing the CMS project. The Project Team leader will liaise with the Program Office on matters of day-to-day management of the project. However, the ultimate management of the CMS project rests with the Program Manager. The essential features of the project management process are:

- Project planning
- Tracking and monitoring
- Progress reporting

6.1. Project Planning

A project plan will be developed to identify:

- Phases and stages
- Start and end dates for each stage
- Tasks
- Fixed milestones for the project (e.g., deadlines for deliverables)
- Approximate schedule for execution of tasks

The plans will be presented in the form of a list containing the above information and a Gantt chart. Plans will be prepared using the MS Project software tool (version 4.0).

In addition to the Project Plan, task sheets will be developed at the start of each phase of the project. The general format of the task sheets will be:

1. Task description
2. Purpose
3. Scope
4. Deliverable
5. Content of deliverable
6. Work method
7. Completion

Upon commencing each new task, team members will review the appropriate task sheet and make appropriate amendments with the approval of the Program Manager. The Program Manager will report task sheet amendments to the subsequent Steering Committee meeting for ratification.

6.2. Tracking and Monitoring

The Program Office will update the CMS project plans prior to each monthly Steering Committee meeting, and provide them to the Program Manager.

The Program Manager will monitor the plans to identify:

- Any slippage to the project
- Additional resource requirements
- Alterations required to the scheduling of tasks

Any findings will be reported to the Steering Committee as part of a progress report.

6.3. Progress Reporting

The main, ongoing planning and control document will be the Progress Report, prepared by the Program Office every two weeks. The report will set out the progress in the previous two weeks, and the plans for the next two weeks. The report will highlight particular problems, especially those relating to time and cost, and will report significant deviations from the overall Project Plan. The Program Office will provide the Progress Reports to the Program Manager for distribution.

Each report will be circulated to all members of the Steering Committee. At each Steering Committee meeting, the previous two Progress Reports will formally be approved.

The general format of the interim progress report is:

1. Introduction
2. Tasks in progress (during the previous two weeks)
3. Tasks completed (during the previous two weeks)
4. Tasks scheduled (during the next two weeks)
5. Other issues

In addition to the Progress Report, the Project Manager will produce an updated Project Plan for each Steering Committee meeting, showing:

- Baseline start and end date of each task
- Actual start and end of each task
- Progress to date in completing each task

The Steering Committee will formally approve the updated plans at each Steering Committee meeting.

6.4. Software Support

Define all standards for documents.

All reports, papers, and textual documentation will be developed using MS Word version 7.0. Other materials will be produced using the following software:

Spreadsheets	MS Excel version 7.0
Graphics and presentations	MS PowerPoint version 7.0
Flowcharts	Visio version 5.0
Project plans	MS Project version 4.0

Requirements for the use of other software should be determined in conjunction with the Project Manager.

6.5. Document Management

6.5.1. Version Control

Each document will start with a version number of 0.1. This version will be incremented by 0.1 each time a new version is released; that is, the second version will be 0.2. Once a document has been approved by the Steering Committee, it will become version 1.0, and will be updated accordingly in the footer of the document. The approved version will be republished only at the explicit request of the Steering Committee. A copy of the updated document will be held in the project files.

Subsequent amendments of the document will incremented by 0.1; for example, 1.1 is the version after the first approved document. Subsequent approvals by the Steering Committee will result in versions 2.0, 3.0, and so on.

Prior to approval by the Steering Committee, a document will bear the description DRAFT in the title of the document and at the center of the footer. Thus, any version other than 1.0, 2.0, 3.0, and so on, will be issued as a draft.

Versions of documents subsequent to that approved by the Steering Committee (i.e., all draft documents after version 1.0) should have changes indicated using revision marks as follows:

- Newly inserted text should be underlined.
- ~~Deleted text should remain in the document but be marked with a strike through.~~
- Where possible, sidebars should be used to highlight the changes.

MS Word 7.0 provides revision-marking facilities that highlight changes automatically. Substantial changes to the document should be described for each new version in the modifications record at the front of the document.

6.5.2. Document Change Control

Once a document has been accepted by the Steering Committee, it will come under formal change control. The purpose of a formal change control is to ensure that changes to formal deliverables are carried out in a controlled manner such that the dependencies between subsequent documents are recognized. Changes to one deliverable can have significant implications for subsequent deliverables, and may involve reworking of some documents.

A request for change can be made by any project participant, and must be made on the Change Request Form, attached as Addendum 2 of this document. The form should be completed and submitted to the Program Manager for authorization and submission to the Steering Committee.

7. CONCLUSION

This plan will be refined and added to during the course of the project as required.

ADDENDUM 1: JOB DESCRIPTIONS

The individual job descriptions will vary according to the structure of your company. Your department head may be responsible for tasks allocated here to the Steering Committee; you may have the Program Manager and Program Office consolidated in a single individual. Adjust tasks and descriptions accordingly.

A. Program Steering Committee

1. Main Purpose of Job

The Program Steering Committee will make the executive decisions required to ensure that the project serves the best interests of the Company, and is consistent with the other projects comprising this program.

2. Position in Organization

The Program Steering Committee has ultimate responsibility for the project.

3. Scope of Job

The Program Steering Committee will make all executive decisions regarding the management, conduct, and control of the project.

4. Responsibilities

The Steering Committee is responsible for:

- Overall management and control of the project
- Approval of project deliverables
- Approval of Progress Reports
- Approval of Project Plans
- Approval of changes to project deliverables

B. Program Manager Job Description

1. Main Purpose of Job

The Program Manager is responsible for overseeing the implementation of the projects on a day-to-day basis.

2. Position in Organization

The Program Manager reports to the Program Steering Committee. The Program Office and the User Project Team Leaders report to the Program Manager.

3. Scope of Job

The Program Manager makes day-to-day decisions about the conduct of the project, but defers to the Steering Committee on issues of:

- Changes to scope
- Changes to cost of the project
- Changes to the overall time frames of the project

The Program Manager may make executive decisions regarding:

- Scheduling of tasks
- Content and method of execution of tasks
- Content of deliverables
- Allocation of resources to tasks

4. Responsibilities

The Program Manager is responsible for developing the Quality Plans, Project Plans, Job and Task Descriptions, companywide communications, and executive briefings.

This includes:

- Managing the Project.
- Converting data.
- Providing the infrastructure (hardware and systems software).
- Providing training and a user help desk.
- Assuring quality.
- Communicating Project information throughout the Company and externally.
- Allocating project tasks to Company staff.
- Presenting Progress Reports, Project Plans and Deliverables, and changes to project deliverables to the Steering Committee.

- Verifying compliance of deliverables with project standards.
- Identifying the appropriate system user Project Team Leaders to carry out reviews of design products.
- Completing review of deliverables by system user Project Team Leaders and Steering Committee.
- Releasing RFPs.
- Providing (recruiting) IS staff.
- Establishing policies and procedures.
- Successfully implementing each of the projects.

C. User Project Team Leaders

1. Main Purpose of Job

The Project Team Leader is responsible for the detailed planning and management of his or her respective project through to a successful conclusion.

2. Position in Organization

Each Team Leader reports to the Program Manager. Each team will be composed of system users, along with IS staff, outside vendors, and a Program Office consultant.

3. Scope of Job

The Team Leader may make day-to-day decisions regarding the conduct of his or her individual project, provided that these decisions do not affect changes to overall scope, cost, and/or schedule. Decisions regarding overall scope, cost, and/or schedule must be brought to the Program Manager for overall project decisions or for subsequent presentation to the Steering Committee.

The Team Leaders may make such recommendations regarding their specific project with regard to:

- Analysis
- Design, in accordance with overall project standards
- RFI/RFP preparation

- Evaluation of vendor proposals and preparation of recommendations for approval by Program Manager and Steering Committee
- User testing and reporting of results
- User training and education
- Test scripts and test data
- Pilot testing
- User acceptance
- Cleaning up user data and documents

4. Responsibilities

Each Team Leader is responsible for developing the detailed work plans, job and task descriptions, and standards for his or her specific project. The Team Leader is also responsible for producing the planning of the project for each stage.

The Team Leader is responsible for selecting the members of his or her specific project team, and for convening meetings until the deliverables, as outlined in subsequent task descriptions, are completed and accepted.

D. Program Office

1. Main Purpose of Job

The Program Office's purpose is to provide support for:

- Managing the program
- Assuring quality
- Building awareness and acceptance of the projects in the program
- Assisting with the implementation of the projects comprising the program

2. Position in Organization

The Program Office reports to the Program Manager.

3. Scope of Job

The Program Office provides resources and expertise to the Program Manager and the user teams.

4. Responsibilities

The Program Office is responsible for:

- Defining the organization and roles of each project participant.
- Monitoring adherence to the organization and roles throughout the project.
- Coordinating all Project Teams to ensure focus on task sequencing, schedule, and effective resource deployment.
- Defining job descriptions for each person working on the project.
- Assisting in the preparation of task sheets that define each task in each project.
- Updating project plans and providing them to the Program Manager.
- Assisting the Program Manager in preparing the Progress Report.
- Monitoring plans to identify potential problems, issues, or slippages; additional resource requirements; changes to scheduling and/or allocation of tasks; changes in risk; and reporting any findings to the Program Manager.
- Preparing high-level designs for each part of the total integrated system.
- Assisting in evaluations of proposals of outside vendors and consultants who join the Project Teams to work on each project.
- Assisting in the development of any RFPs.
- Assisting the Team Leaders of each project to plan and control the project.
- Managing change requests.
- Reporting to the Steering Committee and the Program Manager.
- Building awareness and acceptance of the program throughout the Company and with Company constituencies.
- Reporting to the Program Manager on work in progress and detailed goals.
- Assisting in the preparation of project standards.
- Assisting in assuring quality.

ADDENDUM 2: CHANGE REQUEST FORM

REQUEST FOR CHANGE TO A PROJECT DELIVERABLE

REQUESTOR DETAILS

Serial Number:

Name: ____________________

Date Prepared: ____________

Tel No/Ext: ____________

E-mail: ____________

REQUEST DETAILS

Deliverable to be changed: ________ Current Version: ________

Description of change requested:

Reason for change:

EFFECT OF CHANGE

Design products requiring updating as a result of change:

Estimated effort required:

1 ____________________ ____________ days

2 ____________________ ____________ days

3 ____________________ ____________ days

Total estimated effort required: ________ days

Total estimated cost: $__________

APPROVAL

Program Manager: Date:

Steering Committee (Chair): Date:

Appendix II

Sample Statement of Work (SOW) Between Client and Vendor

Sample Statement of Work (SOW) Between Client and Vendor

This appendix is intended as an example of items to include in the Statement of Work (SOW). Payment milestones should be noted on deliverables, to help in project monitoring. Tasks and dates will vary, of course, according to the specific project. Items shown in italics are variables for you to consider and complete.

Be sure to specify date of contract signing, customer name, and date of this Statement of Work document. Add any other identifying information, such as Client document number, that may help in identifying payments and queries regarding this SOW.

KEY ASSUMPTIONS

The following key assumptions are included in this Statement of Work (SOW).

1. *Locations where the work will be performed.*
2. Vendor will designate a Project Manager, who will be the primary contact for all communications with Client.
3. Client will assign at least one knowledgeable staff person from project initiation, who will work with vendor staff for the duration of the project. The Client personnel assigned to this project

will have the technical/functional/accounting skills necessary to participate in the project.

4. All work is predicated on Vendor's proposal dated 5-01-2001 as submitted to Client. *Specify agreements, such as vendor agreement with Client, approved specifications, and approved Statement of Work, and sequence of documents, so you know which supersedes and overrides the Proposal, in case a discrepancy in functionality or work to be accomplished occurs.* For example, the order of precedence for these documents is that the Vendor Agreement supersedes all other agreements, then the Functional Specifications Document(s), then the Statement of Work and then the Proposal.
5. Any changes to the scope of the project or equipment to be delivered will be managed through the Change Order Procedure. The Change Order Procedure is described in Section 2 of this document.
6. Both parties agree to a reciprocal 48-hour turnaround for response to most business or technical issues. Both parties agree to a reciprocal 96-hour turnaround for response to business or technical issues of a more complex nature.

VENDOR

Vendor will perform the following tasks for Client in accordance with the terms and conditions of the contract:

1. Vendor will provide project leadership services for the tasks in this Statement of Work.
2. Vendor will appoint a Project Manager, who will have explicit responsibility for the administration and technical direction of Vendor's activities. Project leadership duties include:
   - Responsibility for communications with Client
   - Establishment and administration of project leadership procedures
   - Development and implementation of project work plans
   - Measurement and evaluation of project progress against project work plans, budgets, and schedules
   - Report progress of project tasks

- Documentation, maintenance, and update of project issues and their status

TABLE OF CONTENTS

1. TIMELINE

The tasks outlined in this Statement of Work are expected to be finished on or before the dates shown in the project schedule of 7-25-2001.

2. CHANGE ORDER PROCEDURE

The Change Order Procedure applies to all changes to the Agreement dated 6-10-2001 and signed 7-10-2001, the approved Functional Specification Document(s), the approved Statement of Work, and all Change Orders. The Change Order Procedure is defined as follows:

> Any extension or amendment to the Agreement, the Functional Specification Document(s), or the Statement of Work will require both parties to mutually agree to extend or amend these agreements to cover the changed obligations, terms, and conditions. Change orders that are above the amounts allocated for software modification or development, or that alter the scope or the total Agreement amount, will be handled by Vendor as a

separate change order request. Client shall have the right to request that Vendor make modifications, changes, or additions to the program materials. If so requested, Vendor shall evaluate such change order request to determine whether such modification, change, or addition can be provided. All such requests and responses shall be in writing. If Vendor determines that it can provide such modification, change, or addition, Vendor shall prepare a formal change order proposal describing such modification, change, or addition, setting forth any additional charges, efforts required, an estimated time for completion, and any impact upon the existing time schedules. In the event Client desires to alter the scope of work relating to any change order once it is executed by both parties, an additional change order must be executed between Client and Vendor pursuant to the above procedure. With respect to correspondence and approvals of change requests, each party shall have thirty (30) working days to respond to the other, with the exception of complex specifications, which by their nature may require more time.

3. SYSTEM OVERVIEW

In this section, place a brief but complete overview of the system being delivered to the Client.

4. INITIAL SITE VISIT (DELIVERY, INSTALLATION AND CONFIGURATION)—PAYMENT MILESTONE PAYMENT: ($2,000)

Description

This is the first site meeting between the Client representative(s) and the Vendor Project Manager; it includes:

1. Review of application software and interfaces
2. Review of current policies and procedures related to *specify application functions*
3. Identification and documentation of questions, issues, and/or mutual concerns

Roles/Responsibilities

Client:

1. Provide access to central server system.
2. Provide network administrator to assist in the installation.
3. Decide which Client desktops need the application.

Vendor:

1. Install the application software on the central server.
2. Configure individual desktops for running the program.
3. Ensure program runs on all desktops.

Deliverable(s)

1. Application files and one CD-ROM with the application for Information Systems Department

Completion Criteria

1. The task will be completed when the application is loaded and functional on the server and individual desktops in the Client offices.
2. Client will sign off that application is functional.

5. STATEMENT OF WORK

Task Description

The Statement of Work (SOW) is a chronological list of project milestone steps, and is the mutual plan that the project is expected to follow. The SOW will have a Task Description, Roles/Responsibilities, Deliverable(s), and Completion Criteria for each payment milestone and other major tasks.

Roles/Responsibilities

Client:

1. Provide an authorized representative for the review.
2. Review the SOW, and approve, or make, changes.

Vendor:

1. Facilitate the SOW review.
2. Provide SOW draft for review.
3. Secure approval or document changes required.

Deliverable(s)

1. The approved SOW and two signed original approval documents.

Completion Criteria

1. This task will be complete when the Client and Vendor approve the SOW.

6. INVENTORY AND NEEDS ANALYSIS

Task Description

Collect data on department procedures and practices before beginning input of program reference tables. Collect relevant reports for comparison to existing program standards.

Roles/Responsibilities

Client:

1. Provide departmental procedures manual or relevant documents relating to agreement management procedures.
2. Allow access to current agreement management system.
3. Allow access to departmental personnel responsible for agreement management and accounting.
4. Compile current reporting tools for agreement management.

Vendor:

1. Compile all procedures needed for software application management.
2. Review and compare existing agreement management application to data migration information.
3. Compile relevant reports and compare to *vendor product name* system tools.

Deliverable(s)

1. Client will deliver necessary reports for inclusion into the *vendor product name* system.
2. Client will compile any necessary procedural manuals related to agreements.
3. Vendor will provide any necessary enhancements to reports as needed in the *vendor product name* system.

4. Vendor will provide a report summarizing any reports being enhanced or other changes being made.

Completion Criteria

1. The task will be completed following the kick-off meeting and remittance of a report containing an analysis of Client's current reports and procedures during the first meeting.

7. TABLE INPUT (APPLICATION SOFTWARE)—PAYMENT MILESTONE (PAYMENT = $83,750)

Task Description

Review relevant abstract agreement data with *specify appropriate departmental or class of personnel* in order to begin inputting *specify data type.* Train personnel to input into Reference/System tables.

Roles/Responsibilities

Client:

1. *Define Client responsibilities to prepare the data.*
2. *Specify any sequences of activities required.*

Vendor:

1. Work closely in *define assisting tasks here.*
2. *Specify any validation services.*
3. Work with personnel to inventory system variables.

Deliverable(s)

1. Client will complete all input of *specify tables and any sequences of activities.*
2. Ensure reference/system tables are complete.

Completion Criteria

1. All *specify number and types of* tables are complete and correct.
2. Vendor submits a sign-off that tables are complete and correct.

8. DATA INPUT (APPLICATION SOFTWARE)—PAYMENT MILESTONE (PAYMENT = $21,250)

This is similar to previous task in nature, but specifies any input data.

Task Description

Input of *specify data required.*

Roles/Responsibilities

Client:

1. Input of *specify approximate number of datasets* and ancillary information.
2. Input of *approximate number of historical datasets.*
3. Generate sample test data and validation information.

Vendor:

1. Assist in inputting *dataset* information by remote viewing and periodic on-site evaluation.
2. Assist in inputting *specify other* information *(specify approximate number of items)* by remote viewing and periodic on-site evaluation.
3. Assist and train for one month *specify testdata* generation.

Deliverable(s)

1. Vendor will train in *specify areas for training.*
2. Vendor will review *specify datasets* for errors and completeness.
3. Vendor will submit periodic progress reports, indicating what has been, and what remains to be, completed.

Completion Criteria

1. All *specified number and types of data* are inputted and complete.
2. Progress report submitted stating that all have been completed and verified.

9. DATA MIGRATION—PAYMENT MILESTONE (PAYMENT = $10,000)

Task Description

Data migration of *specify types of* data into *specify name of* system.

Roles/Responsibilities

Client:

1. Provide access to *specify any systems to which access is required to perform task.*

Vendor:

1. Migrate *specify datasets, with any qualifications, such as number of fields, number of months, or years of historical data.*

Deliverable(s)

1. *Specify dataset name* migration of data.
2. Report summarizing data migration process and activity.

Completion Criteria

1. *Specify dataset to be verified as complete.*
2. Sign-off by Client that information has been successfully migrated.

10. PROVIDE SYSTEM TRAINING—(PAYMENT MILESTONE PAYMENT 4 @ $3,500 = $14,000 [SEE DELIVERABLES SECTION FOR BREAKOUT])

Task Description

The purpose of this task is to provide operational training for the *vendor system name* users and the system administrators. *If you are having "train the trainer" training, follow that theme throughout the items in this task.*

Roles/Responsibilities

Client:

1. Schedule the users.
2. Provide a list of the users *of Vendor's system.*
3. Provide facilities and basic equipment (chairs, desks, computers, etc.) for the end-user training sessions.

Vendor:

1. Conduct end-user training. *Specify the length of time in days and hours per session that the training will require, the maximum*

number of students per session, and any specifications for classrooms and/or materials reproduction.

2. Conduct system administrator training. *Specify the length of time in days and hours per session that the training will require, the maximum number of students per session, and any specifications for classrooms and/or materials reproduction.*
3. Provide copy of syllabus and agenda. *Specify any time requirements* in advance of training for distribution to attendees.
4. Provide one copy of the end-user training manual for each user to be trained (10 maximum) and one copy in PDF format *or any other materials reproduction needs.*

Deliverable(s)

1. Provide training on program operation at initial installation of program (payment = $3,500).
2. Provide training on *specify datasets and tables* input (payment = $3,500).
3. Provide training on *specify functionality* process (payment = $3,500).
4. Provide final training and user manuals to all end users and system administrators (payment = $3,500).

Completion Criteria

Each task will be complete when the associated deliverable is delivered to Client.

11. CONDUCT PILOT TEST (APPLICATION SOFTWARE)—PAYMENT MILESTONE (PAYMENT = $20,000)

Task Description

Test *specify the items that constitute a systems test, including the length of time system must be run, and any specific functionality to be demonstrated.* This includes prior development of a written test plan and agreement between the parties as to the types and levels of test and acceptance.

Roles and Responsibilities

Client:

1. Select test environment data in conjunction with Vendor.
2. Provide test and acceptance criteria in conjunction with Vendor.

3. Conduct test scenarios in conjunction with Vendor for: *specify each criterion and/function to be tested.*
4. Evaluate pilot test in conjunction with Vendor.
5. Review test plan.

Vendor:

1. Provide assistance in processing test and acceptance criteria in conjunction with Client.
2. Provide test and acceptance criteria in conjunction with Client.
3. Make modifications as needed in conjunction with Vendor.
4. Write test plan.

Deliverable(s)

1. Written test and acceptance plan.
2. Processing pilot test without errors.
3. Client approval of test plan.

Completion Criteria for Pilot Test

1. Vendor signed approval of pilot testing completion.

12. *VENDOR SYSTEM NAME* "LIVE" (SOFTWARE MODIFICATION)—PAYMENT MILESTONE (PAYMENT = $30,000)

Task Description

The purpose of this task is to "advance" the Client from the testing environment to the "live" environment. The "live" environment will enable Vendor to evaluate the structure of the system and make changes as necessary.

Roles/Responsibilities

Client:

1. Provide authorized representative for review and acceptance of the installation.
2. Provide formal list of remaining issues before base system acceptance.

Vendor:

1. Monitor testing environment for any necessary changes.
2. Test system data.
3. Evaluate and correct bug fixes and any change orders.

Deliverable(s)

1. Base system installed and fully operational.
2. Report summarizing all changes made, plus approvals.

Completion Criteria

1. The *base vendor system* is installed and operating on the central file server(s).
2. The Client applications are installed and operating.
3. Client has begun "live" data entry.
4. Changes report is submitted.
5. Sign-off by Client.

13. SYSTEM ACCEPTANCE (DESIGN SERVICES)—PAYMENT MILESTONE (PAYMENT = $15,000)

Task Description

The purpose of this task is to demonstrate to Client that the Vendor system will meet or exceed the requirements of the contract, during a 30-day period. Acceptance of all tests and program operations will be finalized.

Roles/Responsibilities

Client:

1. Utilize the *specify vendor's system name* system.
2. Review and accept the acceptance test plan.
3. Report any discrepancies to Vendor for corrections.
4. Assist in the resolution of any issues.

Vendor:

1. Review and accept the acceptance test plan.
2. Correct any reported program problems within a time period mutually acceptable to both parties.

3. Maintain support communication with Client to ensure proper operation of system for the life of the program.

Deliverable(s)

1. Accepted acceptance test plan.
2. Report summarizing results of testing against acceptance criteria.

Completion Criteria

1. This task will begin immediately after the completion of the prior going-"live" task, and will be complete after (30) consecutive calendar days, during which the software will function per the contract and RFP specifications.
2. Delivery of acceptance test report, summarizing any changes, problems, and their resolutions.
3. Sign-off by Client of acceptance against acceptance plan.

14. ANNUAL MAINTENANCE—PAYMENT MILESTONE (PAYMENT = 3 @ $18,000)

Task Description

Provide annual maintenance of Vendor program, including any upgrades and version changes throughout the periods specified. Also included are program user manual changes and any necessary changes to operate the system within the parameters of the proposal.

Roles/Responsibilities

Client:

1. Utilize the Vendor's system.
2. Maintain the system as instructed by the User's Manual.
3. No custom changes made by the Client.
4. Assist in the resolution of any issues.

Vendor:

1. Provide upgrades and version changes with CD installations.
2. Correct any reported program problems within a time period mutually acceptable to both parties.

3. Maintain support communication with Client to ensure proper operation of *system name* for the life of the program.

Deliverable(s)

This section should specify the length of time and any deliverable during paid maintenance that the client can reasonably anticipate, according to Vendor policy. Media, frequency, and any limitations or restrictions should be specified. For example:

Year 1

1. Provide upgrade/custom and/or version changes with one CD delivered to Client.
2. Provide one color copy and one copy in PDF format of User's Manual changes with every version change.

Completion Criteria

1. Client is fully operational with upgrade/custom and/or version changes for each period described in Deliverables section.

Appendix III

Strengths, Weaknesses, Opportunities, and Threats (SWOT) Analysis

Strengths, Weaknesses, Opportunities, and Threats (SWOT) Analysis

The purpose of a strengths, weaknesses, opportunities, and threats (SWOT) analysis is to identify strategies to implement that may increase your chances of success. This is accomplished by listing the internal and external positive and negative factors affecting the situation. By subsequently identifying strategies for dealing with the negative factors, and for maximizing the positive, you increase your chances for a positive outcome. You can avoid getting into situations where negative factors predominate, and your chances of succeeding are small, and can make better use of resources that might otherwise be wasted.

To accomplish a SWOT analysis, make a four-box grid in which to classify attributes of a company or situation, such as investing effort in preparing a proposal. The classifications are:

Strengths. The positive internal factors, such as a strong and loyal labor force; strong brand-name recognition; innovative new product development; large investment in research and development.

Weaknesses. The negative internal factors, such as poor financial condition, new management, recent incomplete merger.

Opportunities. The positive external factors, such as strong buyer relationships, industry position, advantageous government policy.

Threats. The negative external factors, such as market demographics, declining industry trends, technology trends in the industry, adverse tax law changes, competitor activities.

Use the intersection to identify four sets of strategies to:

- Use your strengths to take advantage of identified external opportunities.
- Use your strengths to overcome external threats.
- Overcome identified weaknesses, and take advantage of opportunities.
- Minimize weaknesses and overcome threats.

Note that the same fact may be a weakness in some situations, while being a definite strength in others.

Glossary

Acceptance Test Typically, the final test on software, which, when successful, indicates completion of the software implementation and reckoning of payments due the developer (or implementor).

Activity A single task of work within a project.

ADM (Arrow Diagramming Method) A method of network diagramming wherein the arrows represent the activities, which are linked in a precedence relationship.

Agreement A contract between two or more parties to deliver a product or service. In government, an agreement is often used for a service-delivery contract, while a contract applies to product delivery.

Application Software Software designed for a particular departmental or functional purpose, for example, word processing or accounting.

Architecture The high-level systems design; that is, the allocation of basic functions across applications, systems software, networks, and hardware.

Bar Chart A scheduling representation where the time span of each activity is shown as a horizontal line, the ends of which correspond to the start and completion of each activity.

Baseline A standard by which things are compared. In a schedule, for example, the baseline would be the first published schedule; as it's updated over time, consecutive schedule iterations would be compared to the original (baseline) to determine how much behind or ahead of schedule the project truly is.

Bean Count A numeric count of the numbers of items in predefined categories that have project importance and implications. The actual detail is not included, however, merely the total of such items.

Brief A (usually) written document summarizing the pertinent information, issues, and decision desired, generally for decision making by the recipient.

Browser A program that accesses and displays information available on the World Wide Web. The most common browsers are Netscape and Internet Explorer.

Burden As in burdened labor rate: The percentage added to actual hourly labor rates in an enterprise, to accommodate the contribution made to overhead and/or administrative costs.

Business Need A requirement that is intrinsic to the existence of an enterprise. For example, a restaurant would need to keep track of its food usage, so that it might price its menu appropriately. A non-business need would be to have the menu printed in blue because it's a pretty color. Typically, systems are best when they satisfy business rather than personal needs.

Change Control The process by which changes to existing or planned software may be recommended, approved, and implemented to software throughout its life.

Change Order A document reflecting recommended and approved changes to existing or planned software, indicating the benefits, resources required to implement, schedule, and costs of the change.

Citrix A company whose technology provides access to server-based applications over networks, client devices, and platforms. The applications are executed on the server, so that the network traffic is reduced, and higher application performance is generally achievable.

Client/Server A network configuration where a computer (server) provides services, such as file access, e-mail routing, or sharing of peripherals, to other computers on the networks (clients). "Fat" clients tend to have lots of programs and data on them, while "thin" clients rely on storage at the server.

COBOL CO(mmon) B(usiness) O(riented) L(anguage) A programming language used for business applications. Used extensively during the days of larger computers, it defined procedures, unlike modern languages, which are object-oriented.

Constraint Something that limits choice of action. In the project management sense, it indicates a restriction in the choices we can make in the areas of costs, schedule, software quality, and risk.

Contingency An amount of resource (time, money, or design margin) inserted into the respective plan (schedule, budget, or performance) to accommodate unexpected occurrences during the project.

Contract Formally documented agreement of parties, specifying the terms and conditions that serve as proof of the obligation between them.

Cost Plus Fixed Fee (CPFF) A contractual form of arrangement whereby the customer reimburses the contractor's actual costs and pays a negotiated fee over and above the costs.

Cost Plus Incentive Fee (CPIF) A contractual form of arrangement whereby the customer reimburses the contractor's actual costs and pays a fee based upon a specified result, such as timely delivery.

COTS (Consumer, Off-the-Shelf) Pre-packaged software, sometimes called "shrink-wrapped" because of its packaging.

CPM (Critical Path Method) A network diagram in which activities are arranged with precedence. The CPM identifies where the schedule has the least amount of flexibility.

Critical Path In a network diagram, the longest path from start to finish, or the path without any slack. This is the path representing the shortest time in which the project can be completed.

Design Freeze The point in a project when no more changes can be made to the design without approval by the change control board after going through the change control approval process.

Drill Down The process of passing through successive layers of information to arrive at the detail necessary to support or explain a higher level of abstraction or summary data.

Dummy Activity An activity that takes no time or resource but is needed for representational purposes, such as connecting activities to indicate a precedence condition.

Earned Value Analysis A performance measurement method that compares the anticipated schedule and budget against the percentage of total budget and percentage of work actually completed, to determine variances from the plan.

Evaluation Matrix A method of comparing products or services according to predetermined prioritized and quantified ratings, to determine the most appropriate candidate.

Executive Committee (or Executive Steering Committee) A subset of the senior executives in an enterprise who are responsible for making the policy decisions on a program or large project level.

Firewall A security scheme that prevents unauthorized users from penetrating a computer network.

Fixed Price A contractual form in which price and fee are predetermined, regardless of the amount of effort and/or cost actually incurred.

Formula-IT The Enterprise Systems Strategy and Implementation Methodology, designed by James E. Kennedy, and used in this book to demonstrate how project management and methodology interrelate. Further information may be obtained by contacting Mr. Kennedy at www.FormulaIT.com

FTP (File Transfer Protocol) A communications protocol governing the transfer of files from one computer to another over a network.

G&A (General & Administrative) Expenses The amount of burden added to labor rates, to reflect the costs incurred for supporting the department, division, or service. There are many formulae for calculating the G&A in enterprises.

Gantt Chart A bar chart named for H. L. Gantt, the industrial engineer who popularized them during World War I.

GroupWise Novell Corporation's e-mail, calendaring, and contact management software product.

Hardware Tangible computer and the associated physical equipment enabling the performance of data-processing or communications functions.

Integration The process of combining subsystems, so that they have a uniform presentation to the user or to other software/hardware components of a system. Typically, the goal of integration is to produce a seamless interface to a user.

Interface As a verb, the way in which systems, their hardware and/or software components, and/or people communicate or interact with one another. As a noun, the point at which any of the components interact with another, such as when a user uses a system.

Internet The worldwide network of computer networks using the TCP/IP network protocol.

Intranet A private internal computer network accessible only by authorized persons, typically members of the organization that owns it. An intranet is separate from a company's Internet site, although it uses the same browser software to view data on the intranet.

ISO 2000 The standard for quality software development and interface as agreed to by the International Organization for Standards (ISO), a worldwide federation of more than 130 national standards bodies. ISO 2000 replaces the prior series, ISO 9000, into a single standard for both quality management and quality assurance.

Issues List (Active, Closed) A list of issues affecting a project that require timely resolution between parties prior to completion.

Lessons Learned A review or document whose purpose is to summarize the experience of a project with things to do or not do in subsequent projects.

Life Cycle In software project management, successive stages in the building of software from concept to completion.

Maintenance The last stage in the software life cycle, during which software may be modified after delivery and acceptance, to improve it, adapt to it changed requirements or environment, or fix bugs that were not found earlier.

Methodology A way of building or delivering software, usually based upon a model, such as life cycle or Waterfall; definition of the technical steps to develop or build the software.

Milestone A significant measurable event in a project, typically the completion of an activity.

MIS (Management Information Systems) Also called IT (information technology) or IS (information systems). Typically, the organizational department or division responsible for implementation and maintenance of the hardware, software, and network resources of an enterprise.

Network A system of computers interconnected to share information. Networks can be LANs (local area networks, typically internal to a department or building, which can be hard-wired) or WANs (wide area networks, which generally use telephone lines, satellite dishes, or radio waves to cover a larger area than a LAN does).

Network Diagram Scheduling tools comprising diagrams that show precedent conditions and sequential constraints for each activity, laid out over time. Common types are PERT, PDM, and ADM.

NT Microsoft's operating system, both server- and client-based.

Object-Oriented Programming Contemporary language/technique that looks at data structures (objects) that have an allowable set of procedures (methods) that can be performed on the object(s) within a given class. This approach, compared to the procedural approach of earlier languages, defines the interfaces between objects tightly, which requires only that the interfaces remain constant, and allows the internal code to be changed.

Obligated Funds Also called committed or encumbered funds, these are the monies already contracted for. Their dollar amounts reflect purchase orders, agreements, and contracts with other individuals and organizations to pay money at some future time.

Oracle A worldwide provider of business-to-business software and services, which include Internet-enabled database, tools, and application products, along with education, consulting, and support services.

Organization Chart A hierarchical diagram representing the lines of authority and responsibility within an organization, such as a project.

Outsource The process of contracting with an external organization to provide services for all or some portion of a task that can be performed internally.

Parametric Cost Estimating A method of estimating using relevant historical data to generate the estimate for a (future) project.

PCAnywhere A software product from Symantec that enables a computer to remotely access other computers' files to perform remote troubleshooting and help desk support over a network, and to access information.

PDM (Precedence Diagramming Method) A form of network diagramming wherein activities are represented on the nodes, which are linked in a precedence relationship.

Pentium Intel's name for its PC chip. The successor to its 80846 chip, it has been called, to date, the Pentium, Pentium II, and Pentium III chips, respectively, to represent faster speeds.

PERT (Program Evaluation and Review Technique) A form of network diagram wherein events are shown as nodes, and connecting arrows indicate the precedence constraints.

Private Sector Generally used to refer to nongovernmental businesses.

Program Multiple projects loosely related by common funding, goals, location, and/or sponsorship, and coordinated. Also refers to software, as in a set of software instructions causing specific functions to be performed in given situations.

Progress Report A report summarizing achievements during a specific time period against preagreed goals, schedule, and budget.

Project An organized undertaking using human and physical resources, performed once, to accomplish a specific goal.

Project Plan Defines the way in which the project will achieve its objectives, using resources, satisfying schedule, goals, performance, risk, and cost constraints; defines how the statement of work will be accomplished. Defines the entire plan for the project, but sometimes is taken only to mean the network diagram.

Proposal A document, often in response to an RFP, submitted to a prospective customer by an organization describing the work the organization offers to do for the prospective customer.

Protocol The rules regarding the formats, sequences, and rules for transferring data between computers. Standard communications protocols include TCP/IP and SNA, for example.

Prototyping The evolutionary delivery of software, allowing user involvement in the design of the software. The user's ability to model pieces of the software during the design can often improve the capability of the delivered product to meet user expectations.

Public Sector Typically refers to governmental and quasi-governmental agencies, as opposed to private or nongovernmental organizations.

Punch List List of items remaining to be done before a project can be considered totally complete.

Quadruple Constraint Expands upon Milton D. Rosenau, Jr.'s Triple Constraint construct by adding risk to the other three objectives a project must satisfy: budget, schedule, and performance.

R&D (Research & Development) An effort whose primary goal is to learn, rather than to develop a finished product to bring to market. The results of an R&D project, however, may be incorporated later in the subsequent product development.

Rapid Application Development (RAD) A software development methodology wherein design continues throughout the project. Particularly well-suited for heavy user involvement.

RFP (Request for Proposal) A document issued by an organization defining work to be done, and requesting the submission of responding proposals in a specified format. The proposals are then evaluated against one another, and the most suitable, based upon the selection criteria delineated in the RFP, is selected.

RFQ (Request for Qualifications) Similar to an RFP, but qualifications are asked for. A slate of candidates satisfying the necessary qualifications may then receive RFPs. Also, **RFQ (Request for Quotation)** Similar to an RFP, except that the desired procurement is for stock items or products. In this case, only price and delivery time need to be responded to.

Risk Any threat to the achievement of project completion according to the project plan.

Risk Management Overseeing the threats to the successful project completion such that these threats are minimized and the Quadruple Constraint is not affected adversely.

Roll Up Typically done on a program level, this involves consolidating multiple project schedules to show an overall program-level schedule. Costs may likewise be "rolled up."

Scope The extent covered by the project, typically in terms of time, general functionality, and to whom provided.

SLA (Service-Level Agreement) An agreement between two organizations, or departments within an organization, to assume specified responsibilities, and to deliver specified services according to specified measurable performance criteria.

SME (Subject Matter Expert) A project participant used in a limited capacity for his or her expertise in a specified area, such as a software product, network configuration, and so on.

SNA (Systems Network Architecture) An IBM-developed communications protocol that ties mainframes together. Developed in the 1970s to provide reliable communications capabilities, SNA works with centrally managed devices rather than remote devices.

Software Project Management Management of the process of developing or delivering software (computer programs).

Sponsor The individual, department, or group funding the project, and under whose aegis the project proceeds.

SQL Server A Microsoft relational database management system (RDBMS) designed for client/server use and accessed by applications using SQL.

Staged Delivery Implementation of software in multiple stages. Although the requirements and design are done for the entire system, the software delivered is usable at each step.

Standards The accepted level used as a measure for developed or delivered software, hardware, procedures, and so on. Standards may exist at many levels—industry, enterprise, departmental.

Statement of Work (SOW) Description of the milestones, or pieces of work, the project is to deliver, and when.

SWOT Analysis The process of identifying the strengths, weaknesses, opportunities, and threats relevant to a given situation to determine future actions.

Systems Software The software or programs used to allocate systemwide resources.

TCP/IP The standard protocol used to transmit data over networks, and as the basis for Internet communications.

Time & Materials (T&M) A contractual form whereby the customer pays the contractor for all time spent and for all materials directly attributable to the project. A fee may be charged as a percentage of all costs incurred if this has been negotiated beforehand.

Triple Constraint The term originated by Milton D. Rosenau, Jr., to describe the three key project objectives that must be simultaneously accomplished: performance specification, time schedule, and monetary budget. This concept is further explored in his books on project management, the most recent of which is *Successful Project Management* (3rd edition, John Wiley & Sons, Inc., 1998). You can also visit his Web site at www.RosenauConsulting.com

UNIX An operating system developed by Bell Labs, and adapted to many different hardware platforms.

Virtual (Project) Team Group of individuals working on a project but from different locations. They communicate, typically through the Internet, as if they were collocated for project purposes and management.

Work Breakdown Structure (WBS) A tree-shaped representation of the work tasks necessary to accomplish project objectives.

Workstation Typically, desktop microcomputer designed for use by a single person. Often used interchangeably with microcomputer, although it generally is more powerful (in capabilities and processing power) than a standalone microcomputer.

World Wide Web The interconnected sites (servers) and files (documents) on the Internet supporting the HyperText Transfer Protocol (HTTP), which supports hypertext and multimedia.

Y2K (Year 2000) Commonly refers to the problems that were anticipated to occur in older program code due to the millennium change, and the risk of those older programs being unable to distinguish 19xx from 20xx because only two digits of date had been included in the older databases or the code.

Bibliography

Bennatan, E. M. *On Time, Within Budget,* 2nd ed. New York: John Wiley & Sons, Inc., 1992.

Frame, J. Davidson. *Managing Projects in Organizations,* revised ed. San Francisco: Jossey-Bass, Inc., 1995.

———. *The New Project Management.* San Francisco: Jossey-Bass, Inc., 1994.

Ginac, Frank P. *Creating High-Performance Software Development Teams.* Upper Saddle River, NJ: Prentice Hall PTR, 2000.

Kemps, Robert R. *Fundamentals of Project Performance Measurement.* Mission Viejo, CA: San Diego Publishing Company, for Humphreys & Associates, Inc., 1992.

Kennedy, James E. *Formula-IT Methodology.* At www.FormulaIT.com, 1998.

King, David. *Project Management Made Simple.* Upper Saddle River, NJ: Prentice Hall PTR, Yourdon Press Computing Series, 1992.

Lewin, Marsha D. *The Overnight Consultant.* New York: John Wiley & Sons, Inc., 1995.

———. *The Consultant's Survival Guide.* New York: John Wiley & Sons, Inc., 1997.

Lewin, Marsha D., and Rosenau, Milton D., Jr. *Software Project Management: Step by Step,* 2nd ed. Los Angeles, CA: Marsha D. Lewin Associates, Inc., 1988.

Lowery, Gwen, and Ferrara, Rob. *Managing Projects with Microsoft Project 98.* New York: John Wiley & Sons, Inc., 1998.

McConnell, Steve. *Software Project Survival Guide.* Redmond, WA: Microsoft Press, 1998.

Ould, Martyn. *Managing Software Quality and Business Risk.* West Sussex, UK: John Wiley & Sons, Inc., 1999.

Perry, William E. *Effective Methods for Software Testing,* 2nd ed. New York: John Wiley & Sons, Inc., 2000.

Project Management Institute. *A Guide to the Project Management Body of Knowledge, PMBOK Guide, 2000 Edition,* Newtown Square, PA: Project Managment Institute, 2000.

Redmill, Felix. *Software Projects Evolutionary vs. Big-Bang Delivery.* West Sussex, UK: John Wiley & Sons, Inc., 1997.

Rosenau, Milton D., Jr. *Successful Project Management,* 3rd ed. New York: John Wiley & Sons, Inc., 1995.

———. (Editor). *The PDMA Handbook of New Product Development.* New York: John Wiley & Sons, Inc., 1996.

———. *Managing the Development of New Products,* New York: John Wiley & Sons, Inc., 1993.

———. *Successful Product Development: Speeding from Opportunity to Profit.* New York: John Wiley & Sons, Inc., 1999.

Whitten, Neal. *Managing Software Development Projects,* 2nd ed. New York: John Wiley & Sons, Inc., 1995.

ARTICLE

Lehman, DeWayne. "Senate: Y2K Fixes Worth the Billions Spent." *Computerworld online,* (March 1, 2000): (www.computerworld.com/cwi/story/0,1199, NAV47_STO41669,00.html)

Index